© 2018
Clement Ampadu
drampadu@hotmail.com

ISBN:978-1-387-60627-6
ID:22576388
www.lulu.com

All rights reserved. No part of this publication may be produced or transmitted in any form or by any means, electronic or mechanical, including photocopying and recording, or in any information storage and retrieval system, without the prior written permission of the author

Contents

Preface   3

Dedication   4

I   Metric and Related Spaces   5

B — Some Results in Dislocated Metric Space   6

6   Common r-Fixed Point Theorem for Higher-Order Ciric Type Contraction in Dislocated Metric Space with Application to Weak Partial Metric Spaces   7
- 6.1   Brief Summary   7
- 6.2   Introduction and Preliminaries   7
- 6.3   Main Result   9
- 6.4   Some Consequences of Main Result   13
- 6.5   Application to Weak Partial Metric Spaces   13
- 6.6   Open Problem I   14

7   Common r-Fixed Point Theorems for Two Pairs of Weakly r-Compatible Mappings under Properties r-(EA) and r-(CLR) in Dislocated Metric Space   16
- 7.1   Brief Summary   16
- 7.2   Introduction and Preliminaries   16
- 7.3   Main Result   18
- 7.4   Some Consequences of Theorem 7.3.4   24
- 7.5   Some Consequences of Theorem 7.3.5   26
- 7.6   Open Problem I   28
- 7.7   Open Problem II   29
- 7.8   Open Problem III   30

8   Common r-Fixed Point Theorems in b-Dislocated Metric Space using r-(E.A) and r-(E.A)-Like Properties   33
- 8.1   Brief Summary   33
- 8.2   Introduction and Preliminaries   33
- 8.3   Main Results   36
- 8.4   Open Problem I   40
- 8.5   Open Problem II   41

9   Common r-Fixed Point Theorems for Mappings in Dislocated Metric Space under r-Compatibility of Type (A)   42
- 9.1   Brief Summary   42
- 9.2   Introduction and Preliminaries   42
- 9.3   Main Results   44
- 9.4   Open Problem I   51
- 9.5   Open Problem II   51

Bibliography   53

Preface

The commuting mapping concept is popular in fixed point theory, and a generalization of it, called compatible mappings appeared in [GERALD JUNGCK, COMPATIBLE MAPPINGS AND COMMON FIXED POINTS, Internat. J. Math. and Math. Sci. Vol. 9 No. 4 (1986) 771-779]. Since the concept of compatible mappings appeared in the literature, new generalizations of it have been found, and the fixed point theory relating these concepts have been investigated. In this book, inspired by higher-order fixed point theory [Clement Ampadu, Fixed Point Theory for Higher-Order Mappings. ISBN: 5800118959925, lulu.com, 2016] we consider when the r-times composition of two mappings commute, and introduce a new concept called r-compatible mappings, we then investigate some higher-order fixed point theory relating r-compatible mappings and related concepts in the setting of dislocated metric spaces.

In Chapter 6, we introduce r-coincidence point of two mappings and weakly r-compatible mappings and obtain the higher-order version of Theorem 2.1 [Fei He, Common fixed point of four self maps on dislocated metric spaces, J. Nonlinear Sci. Appl. 8 (2015), 301-308]. Some consequences with application to weak partial metric spaces are given.

Inspired by the concepts of (EA) property, and for example, see [Amri, M. and El Moutawakil,D. (2002) Some New Common Fixed Point Theorems under Strict Contractive Conditions. Journal of Mathematical Analysis and Applications , 270, 181-188], and (CLR) property, and for example, see [Sintunavarat, W. and Kumam, P. (2011) Common Fixed Points for a Pair of Weakly Compatible Maps in Fuzzy Metric Spaces. Journal of Applied Mathematics , 2011, 1-14]. In Chapter 7 we introduce the properties of r-(EA) and r-(CLR), and obtain the higher-order versions of Theorem 1 and Theorem 3, contained in [Dinesh Panthi, Kumar Subedi, Some Common Fixed Point Theorems for Four Mappings in Dislocated Metric Space, Advances in Pure Mathematics, 2016, 6, 695-712], with some consequences.

The major part of Chapter 8, introduces a concept of r-(EA)-Like property, which is inspired by [K. Wadhwa, H. Dubey and R. Jain, Impact of E. A. Like property on common fixed point theorems in fuzzy metric spaces. J. Adv. Stud. Topology 3 (1) (2012), 52-59], and proves some higher-order common fixed point theorems under this property and property r-(EA).

Finally, in Chapter 9, we revisit the r-compatible mappings of type (A) [Clement Ampadu, Higher-Order Fixed Point Theory in Metric and Multiplicative Metric Space Under r-Compatibility of Mappings and Related Concepts. Lulu.com, 2018. ISBN: 1387503006, 9781387503001], and obtain the higher-order version of Theorem 2 [Dinesh Panthi and P. Sumati Kumari, Common Fixed Point Theorems for Mappings of Compatible Type(A) in Dislocated Metric Space, Nepal Journal of Science and Technology Vol. 16, No.1 (2015) 79-86], with some consequences.

Dedication

This work is dedicated to the nuclear and extended family, friends, colleagues and other people who have shown interest in loving me.

Clement Boateng Ampadu
February 2018

Part I
Metric and Related Spaces

Subpart B

Some Results in Dislocated Metric Space

Chapter 6

Common r-Fixed Point Theorem for Higher-Order Ciric Type Contraction in Dislocated Metric Space with Application to Weak Partial Metric Spaces

6.1 Brief Summary

Abstract

Inspired by higher-order fixed point theory [Clement Ampadu, Fixed Point Theory for Higher-Order Mappings. ISBN: 5800118959925, lulu.com, 2016], we obtain the higher-order version of Theorem 2.1 [Fei He, Common fixed point of four self maps on dislocated metric spaces, J. Nonlinear Sci. Appl. 8 (2015), 301-308], and deduce several corollaries with application to weak partial metric space.

6.2 Introduction and Preliminaries

Definition 6.2.1

[P. Hitzler, A. K. Seda, Dislocated Topologies, J. Electr. Engin., 51 (2000), 3-7] A mapping $d : X \times X \mapsto [0, \infty)$, where X is a nonempty set, is call a dislocated metric (in short, d-metric) if the following conditions holds for all $x, y, z \in X$

(a) $d(x, y) = d(y, x)$

(b) if $d(x, y) = d(y, x) = 0$, then $x = y$

(c) $d(x, y) \leq d(x, z) + d(z, y)$

In particular, we say (X, d) is a dislocated metric space

Definition 6.2.2

[P. Hitzler, A. K. Seda, Dislocated Topologies, J. Electr. Engin., 51 (2000), 3-7] Let (X, d) be a dislocated metric space. Then

(a) A sequence $\{x_n\}$ in X converges to $x \in X$ iff $\lim_{n \to \infty} d(x_n, x) = 0$

(b) A sequence $\{x_n\}$ in X is called a Cauchy sequence iff $\lim_{n,m \to \infty} d(x_n, x_m) = 0$

(c) The space (X, d) is called complete if every Cauchy sequence in X converges in X

The concept of coincidence point of two mappings and weakly compatible mappings have appeared in the literature, and for example, see [Fei He, Common fixed point of four self maps on dislocated metric spaces, J. Nonlinear Sci. Appl. 8 (2015), 301-308]. Now we introduce the following

Definition 6.2.3

Let A, S be self maps of a set X. If for any $r \in \mathbb{N}$, $A^r x = S^r x$ for some $x \in X$, then we say x is a r-coincidence point of A and S. In particular we say the pair A, S are weakly r-compatible if they r-commute at their r-coincidence points.

Definition 6.2.4

Let (X, d) be a dislocated metric space, and let A, B, S, T be four self maps of X. We say the pair (S, T) is a higher-order H-contraction with respect to (A, B) if the following holds for all $x, y \in X$

$$d(S^r x, T^r y) \leq \sum_{q=0}^{r-1} c_q \max\{d(A^{q+1}x, T^{q+1}y), d(B^{q+1}y, S^{q+1}x), d(A^{q+1}x, S^{q+1}x),$$
$$d(B^{q+1}y, T^{q+1}y), 2d(A^{q+1}x, B^{q+1}y)\}$$

where $0 \leq c_q < \frac{1}{2}$ for all $0 \leq q \leq r-1$ and $r \in \mathbb{N}$

Proposition 6.2.5

Let A, S, B, T be four self maps of a dislocated metric space (X, d), where (S, T) is a higher-order H-contraction with respect to (A, B). Put

$$M(x, y) := \max\{d(Ax, Ty), d(By, Sx), d(Ax, Sx),$$
$$d(By, Ty), 2d(Ax, By)\}$$

Now for every pair $x \neq y$, define

$$Z := Z(x, y) = \max_{0 \leq v \leq r-1} \beta^{-v} \frac{d(S^v x, T^v y)}{M(x, y)}$$

then

$$Z = \max_{n \in \mathbb{N} \cup \{0\}} \beta^{-n} \frac{d(S^n x, T^n y)}{M(x, y)}$$

where $\beta \in [0, \frac{1}{2})$

Now combining Proposition 6.2.5 and Definition 6.2.4, we have the following

> **Definition 6.2.6**
>
> Let A, S, B, T be four self maps of a dislocated metric space (X, d). We say (S, T) is a higher-order H-contraction with respect to (A, B), if the following holds for all $x, y \in X$
>
> $$d(S^r x, T^r y) \leq Z\beta^r M(x, y)$$
>
> where
>
> $$M(x, y) := \max\{d(Ax, Ty), d(By, Sx), d(Ax, Sx),$$
> $$d(By, Ty), 2d(Ax, By)\}$$
>
> $Z \geq 1$ is given by the previous Proposition and $\beta \in [0, \frac{1}{2})$

6.3 Main Result

Now our main result is the following

> **Theorem 6.3.1**
>
> Let (X, d) be a dislocated metric space, and let A, B, S, T be four self-maps on X such that, for any $r \in \mathbb{N}$, $T^r(X) \subset A^r(X)$ and $S^r(X) \subset B^r(X)$. Suppose (S, T) is a higher-order H-contraction with respect to (A, B). If for any $r \in \mathbb{N}$, the range of one of the mappings A^r, B^r, S^r, T^r is a complete subspace of X, then
>
> (a) B and T have a r-coincidence point u
>
> (b) A and S have a r-coincidence point v
>
> (c) For any $r \in \mathbb{N}$ $A^r v = S^r v = B^r u = T^r u$
>
> Moreover if the pairs $\{A, S\}$ and $\{B, T\}$ are weakly r-compatible, then A, B, S, T have a unique common r-fixed point y and $d(y, y) = 0$

Proof of Theorem 6.3.1

Let x_0 be an arbitrary point in X. Since $S^r(X) \subseteq B^r(X)$ there exists $x_1 \in X$ such that $B^r x_1 = S^r x_0$. Since $T^r(X) \subseteq A^r(X)$, there exists $x_2 \in X$ such that $A^r x_2 = T^r x_1$. Continuing this process we can construct sequences $\{x_n\}$ and $\{y_n\}$ in X defined by

$$y_{2n} = S^r x_{2n} = B^r x_{2n+1},\ y_{2n+1} = T^r x_{2n+1} = A^r x_{2n+2},\ \text{for all } n \in \mathbb{N}$$

We claim that $\{y_n\}$ is a Cauchy sequence in (X, d). First we prove that for each $n \geq 1$

$$d(y_n, y_{n+1}) \leq \delta d(y_{n-1}, y_n)$$

where $\delta = \max\{\frac{Z\beta^r}{1-Z\beta^r}, 2Z\beta^r\}$. Since (S, T) is a higher-order H-contraction with respect to (A, B), we have

$$d(y_{2n}, y_{2n+1}) = d(S^r x_{2n}, T^r x_{2n+1}) \leq Z\beta^r M(x_{2n}, x_{2n+1})$$

where

$$\begin{aligned}
M(x_{2n}, x_{2n+1}) &= \max\{d(A^r x_{2n}, T^r x_{2n+1}), d(B^r x_{2n+1}, S^r x_{2n}), d(A^r x_{2n}, S^r x_{2n}), \\
&\quad d(B^r x_{2n+1}, T^r x_{2n+1}), 2d(A^r x_{2n}, B^r x_{2n+1})\} \\
&= \max\{d(y_{2n-1}, y_{2n+1}), d(y_{2n}, y_{2n}), d(y_{2n-1}, y_{2n}), d(y_{2n}, y_{2n+1}), \\
&\quad 2d(y_{2n-1}, y_{2n})\} \\
&\leq \max\{d(y_{2n-1}, y_{2n+1}), d(y_{2n}, y_{2n-1}) + d(y_{2n-1}, y_{2n}), d(y_{2n-1}, y_{2n}), \\
&\quad d(y_{2n}, y_{2n+1}), 2d(y_{2n-1}, y_{2n})\} \\
&= \max\{d(y_{2n-1}, y_{2n+1}), 2d(y_{2n-1}, y_{2n}), d(y_{2n}, y_{2n+1})\}
\end{aligned}$$

Now we consider the following three cases

<u>Case I:</u> If $d(y_{2n}, y_{2n+1}) \leq Z\beta^r d(y_{2n-1}, y_{2n+1})$, then it follows that

$$d(y_{2n}, y_{2n+1}) \leq \frac{Z\beta^r}{1 - Z\beta^r} d(y_{2n-1}, y_{2n}) \leq \delta d(y_{2n-1}, y_{2n})$$

<u>Case II:</u> If $d(y_{2n}, y_{2n+1}) \leq 2Z\beta^r d(y_{2n-1}, y_{2n})$, then we deduce that

$$d(y_{2n}, y_{2n+1}) \leq 2Z\beta^r d(y_{2n-1}, y_{2n}) \leq \delta d(y_{2n-1}, y_{2n})$$

<u>Case III:</u> If $d(y_{2n}, y_{2n+1}) \leq Z\beta^r d(y_{2n}, y_{2n+1})$, then since $1 - Z\beta^r \neq 0$, we get that $d(y_{2n}, y_{2n+1}) = 0$, and it follows that $0 = d(y_{2n}, y_{2n+1}) \leq \delta d(y_{2n-1}, y_{2n})$.

Thus, since we also get $d(y_{2n-1}, y_{2n}) \leq \delta d(y_{2n-2}, y_{2n-1})$, then it follows that for each $n \geq 1$, we have, $d(y_n, y_{n+1}) \leq \delta d(y_{n-1}, y_n)$, and by induction we have $d(y_n, y_{n+1}) \leq \delta^n d(y_0, y_1)$. Let $m > n$, then by the triangle inequality, we deduce the following

$$\begin{aligned}
d(y_n, y_m) &\leq d(y_n, y_{n+1}) + d(y_{n+1}, y_{n+2}) + \cdots + d(y_{m-1}, y_m) \\
&\leq (\delta^n + \delta^{n+1} + \cdots + \delta^{m-1}) d(y_0, y_1) \\
&\leq \frac{\delta^n}{1 - \delta} d(y_0, y_1)
\end{aligned}$$

> **Proof of Theorem 6.3.1 Continued**
>
> Since $\delta < 1$, if we take limits in the above as $n, m \to \infty$, we deduce, $\lim_{n,m\to\infty} d(y_n, y_m) = 0$, that is, $\{y_n\}$ is a Cauchy sequence. Consequently, the sub-sequences $\{S^r x_{2n}\}$, $\{B^r x_{2n+1}\}$, $\{T^r x_{2n+1}\}$, and $\{A^r x_{2n+2}\}$ are also Cauchy sequences. Without loss of generality, we may assume that $S^r(X)$ is a complete subspace of X, then $\{S^r x_{2n}\}$ converges to $S^r a$ for some $a \in X$. The definition of $\{y_n\}$ also implies $\{B^r x_{2n+1}\}$, $\{T^r x_{2n+1}\}$, and $\{A^r x_{2n+2}\}$ also converge to $S^r a$. Since $S^r(X) \subset B^r(X)$, there exists $u \in X$ such that $S^r a = B^r u$. By triangle inequality, we have $d(B^r u, B^r u) = d(S^r a, S^r a) \leq d(S^r a, y_n) + d(y_n, S^r a) = 2d(S^r a, y_n)$. Since $\lim_{n\to\infty} d(S^r a, y_n) = 0$, it follows that $d(B^r u, B^r u) = 0$. We claim that $d(B^r u, T^r u) = 0$. Suppose not, that is, suppose $d(B^r u, T^r u) > 0$. Now observe we have the following
>
> $$d(B^r u, T^r u) \leq d(B^r u, S^r x_{2n}) + d(S^r x_{2n}, T^r u)$$
> $$\leq d(B^r u, S^r x_{2n}) + Z\beta^r M(x_{2n}, u)$$
>
> where
>
> $$\begin{aligned} M(x_{2n}, u) &= \max\{d(A^r x_{2n}, T^r u), d(B^r u, S^r x_{2n}), d(A^r x_{2n}, S^r x_{2n}), \\ & \quad d(B^r u, T^r u), 2d(A^r x_{2n}, B^r u)\} \\ &= \max\{d(y_{2n-1}, T^r u), d(B^r u, y_{2n}), d(y_{2n-1}, y_{2n}), d(B^r u, T^r u), \\ & \quad 2d(y_{2n-1}, B^r u)\} \end{aligned}$$
>
> Using the continuity of the d-metric and the fact that $d(B^r u, B^r u) = 0$, if we take limits in the above as $n \to \infty$, we deduce that $d(B^r u, T^r u) \leq Z\beta^r d(B^r u, T^r u)$, and since $1 - Z\beta^r \neq 0$, we get $d(B^r u, T^r u) = 0$, that is, $B^r u = T^r u$, that is, u is a r-coincidence point of B and T. Thus, (a) is proved. Since $T^r(X) \subset A^r(X)$, there is $v \in X$ such that $T^r u = A^r v$. We show that $S^r v = A^r v$. Now observe we have the following
>
> $$d(S^r v, A^r v) = d(S^r v, T^r u) \leq Z\beta^r M(u, v)$$
>
> where
>
> $$\begin{aligned} M(v, u) &= \max\{d(A^r v, T^r u), d(B^r u, S^r v), d(A^r v, S^r v), \\ & \quad d(B^r u, T^r u), 2d(A^r v, B^r u)\} \\ &= \max\{d(B^r u, B^r u), d(A^r v, S^r v), d(A^r v, S^r v), \\ & \quad d(B^r u, B^r u), 2d(B^r u, B^r u)\} \end{aligned}$$
>
> Since $d(B^r u, T^r u) = 0$, it follows that $M(u, v) = d(S^r v, A^r v)$, and thus, $d(S^r v, A^r v) \leq Z\beta^r d(S^r v, A^r v)$. Since $1 - Z\beta^r \neq 0$, we get $d(S^r v, A^r v) = 0$, that is, $S^r v = A^r v$. Thus, v is a r-coincidence point of A and S, and (b) is proved. Since $B^r u = T^r u$, $T^r u = A^r v$, and $A^r v = S^r v$, we see that (c) holds. Now we assume that the pairs $\{A, S\}$ and $\{B, T\}$ are weakly r-compatible. Then,
>
> $$A^r A^r v = A^r S^r v = S^r A^r v = S^r S^r v \qquad B^r B^r u = B^r T^r u = T^r B^r u = T^r T^r u$$

> **Proof of Theorem 6.3.1 Continued**
>
> Let $y = B^r u = T^r u = A^r v = S^r v$. We show that y is a r-fixed point of T. Observe we have the following
> $$d(y, T^r y) = d(S^r v, T^r y) \leq Z\beta^r M(v, y)$$
> where
> $$\begin{aligned} M(v, y) &= \max\{d(A^r v, T^r y), d(B^r y, S^r v), d(A^r v, S^r v), \\ & \quad d(B^r y, T^r y), 2d(A^r v, B^r y)\} \\ &= \max\{d(y, T^r y), d(T^r y, y), d(y, y), \\ & \quad d(T^r y, T^r y), 2d(y, T^r y)\} \end{aligned}$$
>
> Since $d(y, y) = 0$ and $d(T^r y, T^r y) = 0$, from the inequality immediately above, we deduce
> $$d(y, T^r y) \leq 2Z\beta^r d(y, T^r y)$$
> and since $1 - 2Z\beta^r \neq 0$, we deduce from the above inequality that $d(y, T^r y) = 0$, that is, $y = T^r y$. It follows that $B^r y = B^r B^r u = T^r T^r u = T^r y = y$, thus y is a r-fixed point of B. Now observe we have the following
> $$d(y, S^r y) = d(S^r y, T^r y) \leq Z\beta^r M(y, y)$$
> where
> $$\begin{aligned} M(y, y) &= \max\{d(A^r y, T^r y), d(B^r y, S^r y), d(A^r y, S^r y), \\ & \quad d(B^r y, T^r y), 2d(A^r y, B^r y)\} \\ &= \max\{d(y, S^r y), d(S^r y, y), d(y, y), \\ & \quad d(S^r y, S^r y), 2d(y, S^r y)\} \end{aligned}$$
>
> Since $d(y, y) = 0$ and $d(S^r y, S^r y) = 0$, from the inequality immediately above, we deduce
> $$d(y, S^r y) \leq 2Z\beta^r d(y, S^r y)$$
> and since $1 - 2Z\beta^r \neq 0$, we deduce from the above inequality that $d(y, S^r y) = 0$, that is, $y = S^r y$, thus y is a r-fixed point of S. It follows that
> $$A^r y = A^r S^r v = S^r A^r v = S^r y = y$$
> which implies that y is a r-fixed point of A. It now follows that y is a common r-fixed point of S, T, A, B and $d(y, y) = d(y, T^r y) = 0$. Now we show uniqueness. Suppose $u \neq v$ and u, v are two common r-fixed point of S, T, A, B. Now observe we have the following
> $$d(u, v) = d(S^r u, T^r v) \leq Z\beta^r M(u, v)$$
> where
> $$\begin{aligned} M(u, v) &= \max\{d(A^r u, T^r v), d(B^r v, S^r u), d(A^r u, S^r u), \\ & \quad d(B^r v, T^r v), 2d(A^r u, B^r v)\} \\ &= \max\{d(u, v), 2d(u, v)\} \\ &= 2d(u, v) \end{aligned}$$
>
> Thus the inequality immediately above implies $d(u, v) \leq 2Z\beta^r d(u, v)$, and since $1 - 2Z\beta^r \neq 0$, it follows that $d(u, v) = 0$, that is, $u = v$, and the proof is finished.

6.4 Some Consequences of Main Result

Let $A = B$ and $S = T$ in Proposition 6.2.5, and let Z' be the modification on Z in this situation, then we get the following from Theorem 6.3.1

> **Corollary 6.4.1**
>
> Let (X, d) be a d-metric space, and let A, T be two self maps on X such that for any $r \in \mathbb{N}$, $T^r(X) \subset A^r(X)$. Suppose
> $$d(T^r x, T^r y) \leq Z' \beta^r \max\{d(Ax, Ty), d(Ay, Tx), d(Ax, Tx), d(Ay, Ty), 2d(Ax, Ay)$$
> holds for all $x, y \in X$, $\beta \in [0, \frac{1}{2})$, and $r \in \mathbb{N}$. If $A^r(X)$ or $T^r(X)$ is a complete subspace of X, for any $r \in \mathbb{N}$, then A and T have a r-coincidence point in X. Moreover, if the pair $\{A, T\}$ are weakly r-compatible, then A and T have a unique common r-fixed point y in X such that $d(y, y) = 0$

Let I_X be the identity mapping, and let $A = B = I_X$ in Proposition 6.2.5, and let Z'' be the modification on Z in this situation, then we get the following from Theorem 6.3.1

> **Corollary 6.4.2**
>
> Let (X, d) be a d-metric space and let S and T be two self maps on X. Suppose
> $$d(S^r x, T^r y) \leq Z'' \beta^r \max\{d(x, Ty), d(y, Sx), d(x, Sx), d(y, Ty), 2d(x, y)\}$$
> holds for all $x, y \in X$, $\beta \in [0, \frac{1}{2})$, and $r \in \mathbb{N}$. Then S and T have a unique common r-fixed point y in X and $d(y, y) = 0$

6.5 Application to Weak Partial Metric Spaces

> **Definition 6.5.1**
>
> [I. Altun, G. Durmaz, Weak partial metric spaces and some fixed point results, Appl. Gen. Topol., 13 (2012), 179-191; G. Durmaz, O. Acar, I. Altun, Some fixed point results on weak partial metric spaces, Filomat, 27 (2013), 317-326] Let $\rho : X \times X \mapsto [0, \infty)$ be a function where X is a nonempty set. If ρ satisfies the following conditions for all $x, y, z \in X$
>
> (a) $x = y$ iff $\rho(x, x) = \rho(y, y) = \rho(x, y)$
>
> (b) $\rho(x, y) = \rho(y, x)$
>
> (c) $\rho(x, y) \leq \rho(x, z) + \rho(z, y) - \rho(z, z)$
>
> then we say ρ is a weak partial metric on X, and (X, ρ) is called a weak partial metric space.

Definition 6.5.2

[I. Altun, G. Durmaz, Weak partial metric spaces and some fixed point results, Appl. Gen. Topol., 13 (2012), 179-191; G. Durmaz, O. Acar, I. Altun, Some fixed point results on weak partial metric spaces, Filomat, 27 (2013), 317-326] Let (X, ρ) be a weak partial metric space. Then

(a) a sequence $\{x_n\}$ in X converges to $x \in X$ iff $\lim_{n \to \infty} \rho(x_n, x) = \rho(x, x)$

(b) a sequence $\{x_n\}$ in X is called Cauchy iff $\lim_{n,m \to \infty} \rho(x_n, x_m) < \infty$

(c) a sequence $\{x_n\}$ in X is called 0-Cauchy iff $\lim_{n,m \to \infty} \rho(x_n, x_m) = 0$

(d) (X, ρ) is complete if every Cauchy sequence in X converges to a point in X

(e) (X, ρ) is 0-complete if every 0-Cauchy sequence in X converges to a point in X such that $\rho(x, x) = 0$

Since a 0-complete weak partial metric space is a complete dislocated metric space, then from Theorem 6.3.1, we have the following

Corollary 6.5.3

Let (X, ρ) be a weak partial metric space, and let A, B, S, T be four self-maps on X such that, for any $r \in \mathbb{N}$, $T^r(X) \subset A^r(X)$ and $S^r(X) \subset B^r(X)$. Suppose (S, T) is a higher-order H-contraction with respect to (A, B). If for any $r \in \mathbb{N}$, the range of one of the mappings A^r, B^r, S^r, T^r is a complete subspace of X, then

(a) A and S have a r-coincidence point

(b) B and T have a r-coincidence point

Moreover if the pairs $\{A, S\}$ and $\{B, T\}$ are weakly r-compatible, then A, B, S, T have a unique common r-fixed point

6.6 Open Problem I

Definition 6.6.1

Let (X, d) be a dislocated metric space, and let A, B, S, T be four self maps of X. We say the pair (S, T) is a higher-order H-contraction of type II with respect to (A, B) if the following holds for all $x, y \in X$

$$d(S^r x, T^r y) \leq \sum_{q=0}^{r-1} c_q \{ d(A^{q+1}x, T^{q+1}y) + d(B^{q+1}y, S^{q+1}x) + d(A^{q+1}x, S^{q+1}x) +$$
$$d(B^{q+1}y, T^{q+1}y) + d(A^{q+1}x, B^{q+1}y) \}$$

where $0 \leq c_q < \frac{1}{9}$ for all $0 \leq q \leq r-1$ and $r \in \mathbb{N}$

Proposition 6.6.2

Let A, S, B, T be four self maps of a dislocated metric space (X, d), where (S, T) is a higher-order H-contraction of type II with respect to (A, B). Put

$$M(x,y) := \{d(Ax, Ty) + d(By, Sx) + d(Ax, Sx) + \\ d(By, Ty) + d(Ax, By)\}$$

Now for every pair $x \neq y$, define

$$Z := Z(x,y) = \max_{0 \leq v \leq r-1} \beta^{-v} \frac{d(S^v x, T^v y)}{M(x,y)}$$

then

$$Z = \max_{n \in \mathbb{N} \cup \{0\}} \beta^{-n} \frac{d(S^n x, T^n y)}{M(x,y)}$$

where $\beta \in [0, \frac{1}{9})$

Now combining Proposition 6.6.2 and Definition 6.6.1, we have the following

Definition 6.6.3

Let A, S, B, T be four self maps of a dislocated metric space (X, d). We say (S, T) is a higher-order H-contraction of type II with respect to (A, B), if the following holds for all $x, y \in X$

$$d(S^r x, T^r y) \leq Z \beta^r M(x, y)$$

where

$$M(x,y) := \{d(Ax, Ty) + d(By, Sx) + d(Ax, Sx) + \\ d(By, Ty) + d(Ax, By)\}$$

$Z \geq 1$ is given by the previous Proposition and $\beta \in [0, \frac{1}{9})$

Now the open problem is to prove the following

Theorem 6.6.4

Let (X, d) be a dislocated metric space, and let A, B, S, T be four self-maps on X such that, for any $r \in \mathbb{N}$, $T^r(X) \subset A^r(X)$ and $S^r(X) \subset B^r(X)$. Suppose (S, T) is a higher-order H-contraction of type II with respect to (A, B). If for any $r \in \mathbb{N}$, the range of one of the mappings A^r, B^r, S^r, T^r is a complete subspace of X, then

(a) B and T have a r-coincidence point

(b) A and S have a r-coincidence point

Moreover if the pairs $\{A, S\}$ and $\{B, T\}$ are weakly r-compatible, then A, B, S, T have a unique common r-fixed point y and $d(y, y) = 0$

Chapter 7

Common r-Fixed Point Theorems for Two Pairs of Weakly r-Compatible Mappings under Properties r-(EA) and r-(CLR) in Dislocated Metric Space

7.1 Brief Summary

Abstract

Inspired by higher-order fixed point theory [Clement Ampadu, Fixed Point Theory for Higher-Order Mappings. ISBN: 5800118959925, lulu.com, 2016], we obtain the higher-order versions of Theorem 1 and Theorem 3, contained in [Dinesh Panthi, Kumar Subedi, Some Common Fixed Point Theorems for Four Mappings in Dislocated Metric Space, Advances in Pure Mathematics, 2016, 6, 695-712]. Consequently, several Corollaries are deduced.

7.2 Introduction and Preliminaries

Definition 7.2.1

[Hitzler, P. and Seda, A.K. (2000) Dislocated Topologies. Journal of Electrical Engineering, 51, 3-7.] A mapping $d : X \times X \mapsto [0, \infty)$, where X is a nonempty set, is call a dislocated metric (in short, d-metric) if the following conditions holds for all $x, y, z \in X$

(a) $d(x,y) = d(y,x)$

(b) if $d(x,y) = d(y,x) = 0$, then $x = y$

(c) $d(x,y) \leq d(x,z) + d(z,y)$

In particular, we say (X, d) is a dislocated metric space

Definition 7.2.2

[Hitzler, P. and Seda, A.K. (2000) Dislocated Topologies. Journal of Electrical Engineering, 51, 3-7.] A sequence $\{x_n\}$ in a dislocated metric space (X, d) is called Cauchy if for every $\epsilon > 0$, there exists $n_0 \in \mathbb{N}$ such that for all $m, n \geq n_0$, we have $d(x_m, x_n) < \epsilon$

Definition 7.2.3

[Hitzler, P. and Seda, A.K. (2000) Dislocated Topologies. Journal of Electrical Engineering, 51, 3-7.] A sequence in a d-metric space converges with respect to d, if there exists $x \in X$ such that $\lim_{n \to \infty} d(x_n, x) = 0$

Definition 7.2.4

A d-metric space (X, d) is called complete if every Cauchy sequence in it is convergent with respect to d to an element in X

Lemma 7.2.5

Limits in a d-metric space are unique

Let X be a nonempty set. The concept of coincidence point of two self maps of X have appeared in the literature, and for example, see [Dinesh Panthi, Kumar Subedi, Some Common Fixed Point Theorems for Four Mappings in Dislocated Metric Space, Advances in Pure Mathematics, 2016, 6, 695-712]. Now we introduce the following

Definition 7.2.6

Let A and S be two self mappings on a set X. If $A^r x = S^r x$ for some $x \in X$ and any $r \in \mathbb{N}$, then we say x is a r-coincidence point of A and S

The concept of weakly compatible mappings have appeared, for example, see [Jungck, G. and Rhoades, B.E. (1998) Fixed Points for Set Valued Functions without Continuity. Indian Journal of Pure and Applied Mathematics, 29, 227-238]. Now we introduce the following

Definition 7.2.7

Let A and S be self maps of a d-metric space (X, d). We say A and S are r-weakly compatible if they r-commute at their r-coincidence point, that is, $A^r x = S^r x$ for some $x \in X$ and any $r \in \mathbb{N}$ implies $A^r S^r x = S^r A^r x$

The concept of (EA) property have appeared, and for example, see [Amri, M. and El Moutawakil, D. (2002) Some New Common Fixed Point Theorems under Strict Contractive Conditions. Journal of Mathematical Analysis and Applications, 270, 181-188]. Now we introduce the following

Definition 7.2.8

Let A and S be self maps of a d-metric space (X, d). We say the mappings A and S satisfy r-(EA) property, if there is a sequence $\{x_n\} \in X$ such that

$$\lim_{n \to \infty} A^r x_n = \lim_{n \to \infty} S^r x_n = u$$

for some $u \in X$ and any $r \in \mathbb{N}$

The concept of (CLR) property have appeared, and for example, see [Sintunavarat, W. and Kumam, P. (2011) Common Fixed Points for a Pair of Weakly Compatible Maps in Fuzzy Metric Spaces. Journal of Applied Mathematics, 2011, 1-14]. Now we introduce the following

Definition 7.2.9

Let A and S be self maps of a d-metric space (X, d). We say the mappings A and S satisfy r-(CLR) property, if there is a sequence $\{x_n\} \in X$ such that

$$\lim_{n \to \infty} A^r x_n = \lim_{n \to \infty} S^r x_n = A^r u$$

for some $u \in X$ and any $r \in \mathbb{N}$

7.3 Main Result

> **Definition 7.3.1**
>
> Let (X,d) be a dislocated metric space, and let A, B, S, T be four self maps of X. We say the pair (A,B) is a higher-order PS-contraction with respect to (S,T) if the following holds for all $x, y \in X$
>
> $$d(A^r x, B^r y) \leq \sum_{q=0}^{r-1} c_q [d(S^{q+1} y, A^{q+1} x) + d(T^{q+1} x, S^{q+1} y) + d(T^{q+1} x, A^{q+1} x) + \\ d(B^{q+1} y, S^{q+1} y) + d(T^{q+1} x, B^{q+1} y)]$$
>
> where $0 \leq c_q < \frac{1}{8}$ for all $0 \leq q \leq r-1$ and $r \in \mathbb{N}$

> **Proposition 7.3.2**
>
> Let A, S, B, T be four self maps of a dislocated metric space (X, d), where (A, B) is a higher-order PS-contraction with respect to (S, T). Put
>
> $$M(x,y) := [d(Sy, Ax) + d(Tx, Sy) + d(Tx, Ax) + \\ d(By, Sy) + d(Tx, By)]$$
>
> Now for every pair $x \neq y$, define
>
> $$Z := Z(x,y) = \max_{0 \leq v \leq r-1} \beta^{-v} \frac{d(A^v x, B^v y)}{M(x,y)}$$
>
> then
>
> $$Z = \max_{n \in \mathbb{N} \cup \{0\}} \beta^{-n} \frac{d(A^n x, B^n y)}{M(x,y)}$$
>
> where $\beta \in [0, \frac{1}{8})$

Now combining Proposition 7.3.2 and Definition 7.3.1, we have the following

> **Definition 7.3.3**
>
> Let A, S, B, T be four self maps of a dislocated metric space (X, d). We say (A, B) is a higher-order PS-contraction with respect to (S, T), if the following holds for all $x, y \in X$
>
> $$d(A^r x, B^r y) \leq Z \beta^r M(x,y)$$
>
> where
>
> $$M(x,y) := [d(Sy, Ax) + d(Tx, Sy) + d(Tx, Ax) + \\ d(By, Sy) + d(Tx, By)]$$
>
> $Z \geq 1$ is given by the previous Proposition and $\beta \in [0, \frac{1}{8})$

Now our first result concerns a common r-fixed point theorem for two pairs of weakly r-compatible mappings using r-(EA) property

> **Theorem 7.3.4**
>
> Let (X, d) be a dislocated metric space, and A, B, S, T be four self maps of X. Suppose the following conditions hold
>
> (a) $A^r(X) \subseteq S^r(X)$, and $B^r(X) \subseteq T^r(X)$, for any $r \in \mathbb{N}$
>
> (b) (A, B) is a higher-order PS-contraction with respect to (S, T)
>
> (c) the pairs (A, T) or (B, S) satisfy r-(EA) property
>
> (d) the pairs (A, T) and (B, S) are weakly r-compatible
>
> If for any $r \in \mathbb{N}$, $T^r(X)$ is closed, then
>
> (i) The maps A and T have a r-coincidence point
>
> (ii) The maps B and S have a r-coincidence point
>
> (iii) The maps A, B, S, T have a unique common r-fixed point

Proof of Theorem 7.3.4

Assume the pair (A, T) satisfy r-(EA) property, then there exists a sequence $\{x_n\} \in X$ such that $\lim_{n \to \infty} A^r x_n = \lim_{n \to \infty} T^r x_n = u$ for some $u \in X$ and any $r \in \mathbb{N}$. Since $A^r(X) \subseteq S^r(X)$ for any $r \in \mathbb{N}$, there exists a sequence $\{y_n\} \in X$ such that $A^r x_n = S^r y_n$ for any $r \in \mathbb{N}$. Hence

$$\lim_{n \to \infty} A^r x_n = \lim_{n \to \infty} S^r y_n = u$$

Since (A, B) is a higher-order PS-contraction with respect to (S, T) we deduce the following

$$d(A^r x_n, B^r y_n) \leq Z\beta^r [d(S^r y_n, A^r x_n) + d(T^r x_n, S^r y_n) + d(T^r x_n, A^r x_n) + \\ d(B^r y_n, S^r y_n) + d(T^r x_n, B^r y_n)]$$

Since

$$\lim_{n \to \infty} d(T^r x_n, S^r y_n) = \lim_{n \to \infty} d(T^r x_n, A^r x_n) = \lim_{n \to \infty} d(S^r y_n, A^r x_n) = 0$$

and

$$\lim_{n \to \infty} d(B^r y_n, S^r y_n) = \lim_{n \to \infty} d(T^r x_n, B^r y_n) = \lim_{n \to \infty} d(B^r y_n, u)$$

If we take limits in the above inequality, we deduce that

$$\lim_{n \to \infty} d(B^r y_n, u) \leq 2Z\beta^r \lim_{n \to \infty} d(B^r y_n, u)$$

and since $1 - 2Z\beta^r \neq 0$, we deduce that $\lim_{n \to \infty} d(B^r y_n, u) = 0$, that is, $\lim_{n \to \infty} B^r y_n = u$. It now follows that

$$\lim_{n \to \infty} A^r x_n = \lim_{n \to \infty} T^r x_n = \lim_{n \to \infty} B^r y_n = \lim_{n \to \infty} S^r y_n = u$$

Assume $T^x(X)$ is closed for any $r \in \mathbb{N}$, then there exists $v \in X$ such that $T^r v = u$. We show that $A^r v = u$. Since (A, B) is a higher-order PS-contraction with respect to (S, T) we deduce the following

$$d(A^r v, B^r y_n) \leq Z\beta^r [d(S^r y_n, A^r v) + d(T^r v, S^r y_n) + d(T^r v, A^r v) + \\ d(B^r y_n, S^r y_n) + d(T^r v, B^r y_n)]$$

Since

$$\lim_{n \to \infty} d(T^r v, S^r y_n) = \lim_{n \to \infty} d(B^r y_n, S^r y_n) = \lim_{n \to \infty} d(T^r v, B^r y_n) = 0$$

and

$$\lim_{n \to \infty} d(S^r y_n, A^r v) = d(u, A^r v)$$

If we take limits in the inequality immediately above, we deduce that

$$d(A^r v, u) \leq 2Z\beta^r d(u, A^r v)$$

and since $1 - Z\beta^r \neq 0$, the above inequality implies $d(u, A^r v) = 0$, that is, $u = A^r v$. It now follows that $A^r v = u = T^r v$. Thus, v is the r-coincidence point of (A, T). Since $A^r(X) \subseteq S^r(X)$, there exists $w \in X$ such that $A^r v = S^r w = u$. We now show that $B^r w = u$.

> **Proof of Theorem 7.3.4 Continued**
>
> Since (A, B) is a higher-order PS-contraction with respect to (S, T) we deduce the following
>
> $$\begin{aligned} d(u, B^r w) &\leq Z\beta^r [d(S^r w, A^r v) + d(T^r v, S^r w) + d(T^r v, A^r v) + \\ &\quad d(B^r w, S^r w) + d(T^r v, B^r w)] \\ &= Z\beta^r [3d(u, u) + 2d(B^r w, u)] \\ &\leq 8Z\beta^r d(u, B^r w) \end{aligned}$$
>
> Since $1 - 8Z\beta^r \neq 0$, the above inequality implies $d(u, B^r w) = 0$, in other words, $u = B^r w$. It follows that $B^r w = u = S^r w$. It follows that w is the r-coincidence point of the maps B and S. Now we have
>
> $$u = B^r w = S^r w = T^r v = A^r v$$
>
> Since the pairs (B, S) and (A, T) are weakly r-compatible, it follows that
>
> $$B^r S^r w = S^r B^r w$$
> $$T^r A^r v = A^r T^r v$$
> $$T^r u = T^r A^r v = A^r T^r v = A^r u$$
> $$S^r u = B^r S^r w = S^r B^r w = B^r u$$
>
> We now show that $B^r u = u$. Since (A, B) is a higher-order PS-contraction with respect to (S, T) we deduce the following
>
> $$\begin{aligned} d(u, B^r u) = d(A^r v, B^r u) &\leq Z\beta^r [d(S^r u, A^r v) + d(T^r v, S^r u) + d(T^r v, A^r v) + \\ &\quad d(B^r u, S^r u) + d(T^r v, B^r u)] \\ &= Z\beta^r [3d(u, B^r u) + d(u, u) + d(B^r u, B^r u)] \\ &\leq 7Z\beta^r d(u, B^r u) \end{aligned}$$
>
> Since $1 - 7Z\beta^r \neq 0$, the above inequality implies $d(u, B^r u) = 0$, that is, $B^r u = u$. Thus $u = B^r u = S^r u$. Similarly, we have, $A^r u = u = T^r u$. Hence,
>
> $$u = A^r u = B^r u = S^r u = T^r u$$
>
> It follows that u is the common r-fixed point of A, B, S, T. For uniqueness, suppose z is another common r-fixed point of A, B, S, T, and $z \neq u$. Since (A, B) is a higher-order PS-contraction with respect to (S, T) we deduce the following
>
> $$\begin{aligned} d(u, z) = d(A^r u, B^r z) &\leq Z\beta^r [d(S^r z, A^r u) + d(T^r u, S^r z) + d(T^r u, A^r u) + \\ &\quad d(B^r z, S^r z) + d(T^r u, B^r z)] \\ &= Z\beta^r [3d(u, z) + d(u, u) + d(z, z)] \\ &\leq 7Z\beta^r d(u, z) \end{aligned}$$
>
> Since $1 - 7Z\beta^r \neq 0$, the above inequality implies $d(u, z) = 0$, that is, $z = u$. It follows that the common r-fixed point is unique, and the proof is finished.

Now our next result concerns a common r-fixed point theorem for weakly r-compatible mappings using r-(CLR) property

> **Theorem 7.3.5**
>
> Let (X, d) be a dislocated metric space, and A, B, S, T be four self maps of X. Suppose the following conditions hold
>
> (a) $A^r(X) \subseteq S^r(X)$, and $B^r(X) \subseteq T^r(X)$, for any $r \in \mathbb{N}$
>
> (b) (A, B) is a higher-order PS-contraction with respect to (S, T)
>
> (c) the pairs (A, T) or (B, S) satisfy r-(CLR) property
>
> (d) the pairs (A, T) and (B, S) are weakly r-compatible
>
> Then
>
> (i) The maps A and T have a r-coincidence point
>
> (ii) The maps B and S have a r-coincidence point
>
> (iii) The maps A, B, S, T have a unique common r-fixed point

Proof of Theorem 7.3.5

Assume that the pair (A, T) satisfy r-(CLR) property, then there exists a sequence $\{x_n\} \in X$ such that
$$\lim_{n \to \infty} A^r x_n = \lim_{n \to \infty} T^r x_n = A^r x$$
for some $x \in X$. Since for any $r \in \mathbb{N}$, $A^r(X) \subseteq S^r(X)$, there exists a sequence $\{y_n\} \in X$ such that $\lim_{n \to \infty} A^r x_n = \lim_{n \to \infty} S^r y_n = A^r x$. Now we show that $\lim_{n \to \infty} B^r y_n = A^r x$. Since (A, B) is a higher-order PS-contraction with respect to (S, T) we deduce the following

$$d(A^r x_n, B^r y_n) \leq Z\beta^r [d(T^r x_n, S^r y_n) + d(T^r x_n, A^r x_n) + d(B^r y_n, S^r y_n) \\ + d(T^r x_n, B^r y_n) + d(S^r y_n, A^r x_n)]$$

Since
$$\lim_{n \to \infty} d(T^r x_n, S^r y_n) = \lim_{n \to \infty} d(T^r x_n, A^r x_n) = \lim_{n \to \infty} d(S^r y_n, A^r x_n) = 0$$
and
$$\lim_{n \to \infty} d(A^r x_n, B^r y_n) = \lim_{n \to \infty} d(B^r y_n, S^r y_n) = \lim_{n \to \infty} d(A^r x, B^r y_n)$$

If we take limits in the inequality immediately above we get
$$\lim_{n \to \infty} d(A^r x, B^r y_n) \leq 2Z\beta^r \lim_{n \to \infty} d(A^r x, B^r y_n)$$

and since $1 - 2Z\beta^r \neq 0$, we get $\lim_{n \to \infty} d(A^r x, B^r y_n) = 0$, that is, $\lim_{n \to \infty} B^r y_n = A^r x$. It now follows that
$$\lim_{n \to \infty} A^r x_n = \lim_{n \to \infty} T^r x_n = \lim_{n \to \infty} B^r y_n = \lim_{n \to \infty} S^r y_n = A^r x$$

Assume $A^r \subseteq S^r(X)$, then there exists $v \in X$ such that $A^r x = S^r v$. We now show that $B^r v = S^r v$. Since (A, B) is a higher-order PS-contraction with respect to (S, T) we deduce the following

$$d(A^r x_n, B^r v) \leq Z\beta^r [d(T^r x_n, S^r v) + d(T^r x_n, A^r x_n) + d(B^r v, S^r v) \\ + d(T^r x_n, B^r v) + d(S^r v, A^r x_n)]$$

Since
$$\lim_{n \to \infty} d(T^r x_n, B^r v) = d(A^r x, B^r v) = d(S^r v, B^r v)$$
and
$$\lim_{n \to \infty} d(T^r x_n, S^r v) = \lim_{n \to \infty} d(T^r x_n, A^r x_n) = \lim_{n \to \infty} d(S^r v, A^r x_n) = 0$$

If we take limits in the inequality immediately above, we deduce that
$$d(S^r v, B^r v) \leq 2Z\beta^r d(S^r v, B^r v)$$

and since $1 - 2Z\beta^r \neq 0$, the above inequality implies that $d(S^r v, B^r v) = 0$, and thus, $S^r v = B^r v$. It now follows that v is the r-coincidence point of the maps B and S. Now we have
$$S^r v = B^r v = A^r x = w \, (say)$$

Since the pair (B, S) is weakly r-compatible, it follows that
$$B^r S^r v = S^r B^r v \implies B^r w = S^r w$$

> **Proof of Theorem 7.3.5 Continued**
>
> Since $B^r(X) \subseteq T^r(X)$, there exists a point $u \in X$ such that $B^r v = T^r u$. We claim that $T^r u = A^r u = w$. Since (A, B) is a higher-order PS-contraction with respect to (S, T) we deduce the following
>
> $$\begin{aligned} d(A^r u, B^r v) &\leq Z\beta^r [d(T^r u, S^r v) + d(T^r u, A^r u) + d(B^r v, S^r v) \\ &\quad + d(T^r u, B^r v) + d(S^r v, A^r u)] \\ &= Z\beta^r [d(B^r v, B^r v) + d(B^r v, A^r u) + d(B^r v, B^r v) \\ &\quad + d(B^r v, B^r v) + d(B^r v, A^r u)] \\ &= Z\beta^r [3d(B^r v, B^r v) + 2d(B^r v, A^r u)] \\ &\leq 8Z\beta^r d(B^r v, A^r u) \end{aligned}$$
>
> Since $1 - 8Z\beta^r \neq 0$, the above inequality implies $d(A^r u, B^r v) = 0$, that is,
>
> $$A^r u = B^r v$$
>
> It now follows that $A^r u = B^r v = T^r u = w$. In particular, u is the r-coincidence point of A and T. Since the pair (A, T) is weakly r-compatible, then,
>
> $$A^r T^r u = T^r A^r u \implies A^r w = T^r w$$
>
> We now show $A^r w = w$. Since (A, B) is a higher-order PS-contraction with respect to (S, T) we deduce the following
>
> $$\begin{aligned} d(A^r w, w) &= d(A^r w, B^r v) \\ &\leq Z\beta^r [d(T^r w, S^r v) + d(T^r w, A^r w) + d(B^r v, S^r v) \\ &\quad + d(T^r w, B^r v) + d(S^r v, A^r w)] \\ &= Z\beta^r [3d(A^r w, w) + d(A^r w, A^r w) + d(w, w)] \\ &\leq 7\beta^r d(A^r w, w) \end{aligned}$$
>
> Since $1 - 7Z\beta^r \neq 0$, it follows that $d(A^r w, w) = 0$, that is, $A^r w = w$. Similarly, we obtain $B^r w = w$. It now follows that
>
> $$A^r w = B^r w = S^r w = T^r w = w$$
>
> Hence w is the common r-fixed point of A, B, S, T. For uniqueness, suppose $z \neq w$ is another common r-fixed point of A, B, S, T. Since (A, B) is a higher-order PS-contraction with respect to (S, T) we deduce the following
>
> $$\begin{aligned} d(w, z) &= d(A^r w, B^r z) \\ &\leq Z\beta^r [d(T^r w, S^r z) + d(T^r w, A^r w) + d(B^r z, S^r z) \\ &\quad + d(T^r w, B^r z) + d(S^r z, A^r w)] \\ &= Z\beta^r [3d(w, z) + d(w, w) + d(z, z)] \\ &\leq 7\beta^r d(w, z) \end{aligned}$$
>
> Since $1 - 7Z\beta^r \neq 0$, it follows from the above inequality that $d(w, z) = 0$, that is, $w = z$, and the proof is finished.

7.4 Some Consequences of Theorem 7.3.4

If $A = B$ in Theorem 7.3.4, then we get the following

> **Corollary 7.4.1**
>
> Let (X,d) be a dislocated metric space, and A, S, T be three self maps of X. Suppose the following conditions hold
>
> (a) $A^r(X) \subseteq S^r(X)$, and $A^r(X) \subseteq T^r(X)$, for any $r \in \mathbb{N}$
>
> (b) A is a higher-order PS-contraction with respect to (S,T), that is,
>
> $$d(A^r x, A^r y) \leq Z' \beta^r Q(x,y)$$
>
> where
>
> $$Q(x,y) := [d(Sy, Ax) + d(Tx, Sy) + d(Tx, Ax) + \\ d(Ay, Sy) + d(Tx, Ay)]$$
>
> $Z' \geq 1$ is modified from Proposition 2.2 and $\beta \in [0, \frac{1}{8})$
>
> (c) the pairs (A,T) or (A,S) satisfy r-(EA) property
>
> (d) the pairs (A,T) and (A,S) are weakly r-compatible
>
> If for any $r \in \mathbb{N}$, $T^r(X)$ is closed, then
>
> (i) The maps A and T have a r-coincidence point
>
> (ii) The maps A and S have a r-coincidence point
>
> (iii) The maps A, S, T have a unique common r-fixed point

If $S = T$ in Theorem 7.3.4, then we get the following

> **Corollary 7.4.2**
>
> Let (X,d) be a dislocated metric space, and A, B, S be three self maps of X. Suppose the following conditions hold
>
> (a) $A^r(X) \subseteq S^r(X)$, and $B^r(X) \subseteq S^r(X)$, for any $r \in \mathbb{N}$
>
> (b) (A,B) is a higher-order PS-contraction with respect to S, that is,
>
> $$d(A^r x, B^r y) \leq Z'' \beta^r R(x,y)$$
>
> where
>
> $$R(x,y) := [d(Sy, Ax) + d(Sx, Sy) + d(Sx, Ax) + \\ d(By, Sy) + d(Sx, By)]$$
>
> $Z'' \geq 1$ is modified from Proposition 2.2 and $\beta \in [0, \frac{1}{8})$
>
> (c) the pairs (A,S) or (B,S) satisfy r-(EA) property
>
> (d) the pairs (A,S) and (B,S) are weakly r-compatible
>
> If for any $r \in \mathbb{N}$, $S^r(X)$ is closed, then
>
> (i) The maps A and S have a r-coincidence point
>
> (ii) The maps B and S have a r-coincidence point
>
> (iii) The maps A, B, S have a unique common r-fixed point

If $A = B$ and $S = T$ in Theorem 7.3.4, then we get the following

> **Corollary 7.4.3**
>
> Let (X,d) be a dislocated metric space, and A, S be two self maps of X. Suppose the following conditions hold
>
> (a) $A^r(X) \subseteq S^r(X)$, for any $r \in \mathbb{N}$
>
> (b) A is a higher-order PS-contraction with respect to S, that is,
>
> $$d(A^r x, A^r y) \leq Z''' \beta^r K(x,y)$$
>
> where
>
> $$K(x,y) := [d(Sy, Ax) + d(Sx, Sy) + d(Sx, Ax) + \\ d(Ay, Sy) + d(Sx, Ay)]$$
>
> $Z''' \geq 1$ is modified from Proposition 2.2 and $\beta \in [0, \frac{1}{8})$
>
> (c) the pair (A, S) satisfy r-(EA) property
>
> (d) the pair (A, S) is weakly r-compatible
>
> If for any $r \in \mathbb{N}$, $S^r(X)$ is closed, then A and S have a unique common r-fixed point

7.5 Some Consequences of Theorem 7.3.5

If $A = B$ in Theorem 7.3.5, then we get the following

> **Corollary 7.5.1**
>
> Let (X, d) be a dislocated metric space, and A, S, T be three self maps of X. Suppose the following conditions hold
>
> (a) $A^r(X) \subseteq S^r(X)$, and $A^r(X) \subseteq T^r(X)$, for any $r \in \mathbb{N}$
>
> (b) A is a higher-order PS-contraction with respect to (S, T), that is,
>
> $$d(A^r x, A^r y) \leq Z' \beta^r Q(x, y)$$
>
> where
>
> $$Q(x, y) := [d(Sy, Ax) + d(Tx, Sy) + d(Tx, Ax) + \\ d(Ay, Sy) + d(Tx, Ay)]$$
>
> $Z' \geq 1$ is modified from Proposition 2.2 and $\beta \in [0, \frac{1}{8})$
>
> (c) the pairs (A, T) and (A, S) satisfy r-(CLR) property
>
> (d) the pairs (A, T) and (A, S) are weakly r-compatible
>
> Then
>
> (i) The maps A and T have a r-coincidence point
>
> (ii) The maps A and S have a r-coincidence point
>
> (iii) The maps A, S, T have a unique common r-fixed point

If $S = T$ in Theorem 7.3.5, then we get the following

Corollary 7.5.2

Let (X, d) be a dislocated metric space, and A, B, S be three self maps of X. Suppose the following conditions hold

(a) $A^r(X) \subseteq S^r(X)$, and $B^r(X) \subseteq S^r(X)$, for any $r \in \mathbb{N}$

(b) (A, B) is a higher-order PS-contraction with respect to S, that is,
$$d(A^r x, B^r y) \leq Z'' \beta^r R(x, y)$$
where
$$R(x, y) := [d(Sy, Ax) + d(Sx, Sy) + d(Sx, Ax) + \\ d(By, Sy) + d(Sx, By)]$$

$Z'' \geq 1$ is modified from Proposition 2.2 and $\beta \in [0, \frac{1}{8})$

(c) the pairs (A, S) or (B, S) satisfy r-(CLR) property

(d) the pairs (A, S) and (B, S) are weakly r-compatible

Then

(i) The maps A and S have a r-coincidence point

(ii) The maps B and S have a r-coincidence point

(iii) The maps A, B, S have a unique common r-fixed point

If $A = B$ and $S = T$ in Theorem 7.3.5, then we get the following

Corollary 7.5.3

Let (X, d) be a dislocated metric space, and A, S be two self maps of X. Suppose the following conditions hold

(a) $A^r(X) \subseteq S^r(X)$, for any $r \in \mathbb{N}$

(b) A is a higher-order PS-contraction with respect to S, that is,
$$d(A^r x, A^r y) \leq Z''' \beta^r K(x, y)$$
where
$$K(x, y) := [d(Sy, Ax) + d(Sx, Sy) + d(Sx, Ax) + \\ d(Ay, Sy) + d(Sx, Ay)]$$

$Z''' \geq 1$ is modified from Proposition 2.2 and $\beta \in [0, \frac{1}{8})$

(c) the pair (A, S) satisfy r-(CLR) property

(d) the pair (A, S) is weakly r-compatible

Then

(i) The maps A and S have a r-coincidence point

(ii) The maps A and S have a unique r-common fixed point

7.6 Open Problem I

Definition 7.6.1

Let (X, d) be a dislocated metric space, and let A, B, S, T be four self maps of X. We say the pair (A, B) is a higher-order PS-contraction of type II with respect to (S, T) if the following holds for all $x, y \in X$

$$d(A^r x, B^r y) \leq \sum_{q=0}^{r-1} c_q \max\{d(S^{q+1}y, A^{q+1}x), d(T^{q+1}x, S^{q+1}y), d(T^{q+1}x, A^{q+1}x),$$
$$d(B^{q+1}y, S^{q+1}y), d(T^{q+1}x, B^{q+1}y)\}$$

where $0 \leq c_q < \frac{1}{2}$ for all $0 \leq q \leq r-1$ and $r \in \mathbb{N}$

Proposition 7.6.2

Let A, S, B, T be four self maps of a dislocated metric space (X, d), where (A, B) is a higher-order PS-contraction of type II with respect to (S, T). Put

$$M(x, y) := \max\{d(Sy, Ax), d(Tx, Sy), d(Tx, Ax),$$
$$d(By, Sy), d(Tx, By)\}$$

Now for every pair $x \neq y$, define

$$Z := Z(x, y) = \max_{0 \leq v \leq r-1} \beta^{-v} \frac{d(A^v x, B^v y)}{M(x, y)}$$

then

$$Z = \max_{n \in \mathbb{N} \cup \{0\}} \beta^{-n} \frac{d(A^n x, B^n y)}{M(x, y)}$$

where $\beta \in [0, \frac{1}{2})$.

Now combining Proposition 7.6.2 and Definition 7.6.1, we have the following

Definition 7.6.3

Let A, S, B, T be four self maps of a dislocated metric space (X, d). We say (A, B) is a higher-order PS-contraction of type II with respect to (S, T), if the following holds for all $x, y \in X$

$$d(A^r x, B^r y) \leq Z \beta^r M(x, y)$$

where

$$M(x, y) := \max\{d(Sy, Ax), d(Tx, Sy), d(Tx, Ax),$$
$$d(By, Sy), d(Tx, By)\}$$

$Z \geq 1$ is given by the previous Proposition and $\beta \in [0, \frac{1}{2})$

Now the open problem is as follows

> **Theorem 7.6.4**
>
> Let (X,d) be a dislocated metric space, and A, B, S, T be four self maps of X. Suppose the following conditions hold
>
> (a) $A^r(X) \subseteq S^r(X)$, and $B^r(X) \subseteq T^r(X)$, for any $r \in \mathbb{N}$
>
> (b) (A, B) is a higher-order PS-contraction of type II with respect to (S, T)
>
> (c) the pairs (A, T) or (B, S) satisfy r-(EA) property
>
> (d) the pairs (A, T) and (B, S) are weakly r-compatible
>
> If for any $r \in \mathbb{N}$, $T^r(X)$ is closed, then
>
> (i) The maps A and T have a r-coincidence point
>
> (ii) The maps B and S have a r-coincidence point
>
> (iii) The maps A, B, S, T have a unique common r-fixed point

7.7 Open Problem II

Prove that if $A = B$ in Theorem 7.6.4, then we get the following

> **Corollary 7.7.1**
>
> Let (X, d) be a dislocated metric space, and A, S, T be three self maps of X. Suppose the following conditions hold
>
> (a) $A^r(X) \subseteq S^r(X)$, and $A^r(X) \subseteq T^r(X)$, for any $r \in \mathbb{N}$
>
> (b) A is a higher-order PS-contraction of type II with respect to (S, T), that is,
>
> $$d(A^r x, A^r y) \leq Z'\beta^r Q(x, y)$$
>
> where
>
> $$Q(x, y) := \max\{d(Sy, Ax), d(Tx, Sy), d(Tx, Ax),$$
> $$d(Ay, Sy), d(Tx, Ay)\}$$
>
> $Z' \geq 1$ is modified from Proposition 7.6.2 and $\beta \in [0, \frac{1}{2})$
>
> (c) the pairs (A, T) or (A, S) satisfy r-(EA) property
>
> (d) the pairs (A, T) and (A, S) are weakly r-compatible
>
> If for any $r \in \mathbb{N}$, $T^r(X)$ is closed, then
>
> (i) The maps A and T have a r-coincidence point
>
> (ii) The maps A and S have a r-coincidence point
>
> (iii) The maps A, S, T have a unique common r-fixed point

Prove that if $S = T$ in Theorem 7.6.4, then we get the following

> **Corollary 7.7.2**
>
> Let (X,d) be a dislocated metric space, and A, B, S be three self maps of X. Suppose the following conditions hold
>
> (a) $A^r(X) \subseteq S^r(X)$, and $B^r(X) \subseteq S^r(X)$, for any $r \in \mathbb{N}$
>
> (b) (A, B) is a higher-order PS-contraction of type II with respect to S, that is,
> $$d(A^r x, B^r y) \leq Z'' \beta^r R(x,y)$$
> where
> $$R(x,y) := \max\{d(Sy, Ax), d(Sx, Sy), d(Sx, Ax),$$
> $$d(By, Sy), d(Sx, By)\}$$
> $Z'' \geq 1$ is modified from Proposition 7.6.2 and $\beta \in [0, \frac{1}{2})$
>
> (c) the pairs (A, S) or (B, S) satisfy r-(EA) property
>
> (d) the pairs (A, S) and (B, S) are weakly r-compatible
>
> If for any $r \in \mathbb{N}$, $S^r(X)$ is closed, then
>
> (i) The maps A and S have a r-coincidence point
>
> (ii) The maps B and S have a r-coincidence point
>
> (iii) The maps A, B, S have a unique common r-fixed point

Prove that if $A = B$ and $S = T$ in Theorem 7.6.4, then we get the following

> **Corollary 7.7.3**
>
> Let (X,d) be a dislocated metric space, and A, S be two self maps of X. Suppose the following conditions hold
>
> (a) $A^r(X) \subseteq S^r(X)$, for any $r \in \mathbb{N}$
>
> (b) A is a higher-order PS-contraction of type II with respect to S, that is,
> $$d(A^r x, A^r y) \leq Z''' \beta^r K(x,y)$$
> where
> $$K(x,y) := \max\{d(Sy, Ax), d(Sx, Sy), d(Sx, Ax),$$
> $$d(Ay, Sy), d(Sx, Ay)\}$$
> $Z''' \geq 1$ is modified from Proposition 7.6.2 and $\beta \in [0, \frac{1}{2})$
>
> (c) the pair (A, S) satisfy r-(EA) property
>
> (d) the pair (A, S) is weakly r-compatible
>
> If for any $r \in \mathbb{N}$, $S^r(X)$ is closed, then A and S have a unique common r-fixed point

7.8 Open Problem III

Prove the following which concerns a common r-fixed point theorem for weakly r-compatible mappings using r-(CLR) property

> **Theorem 7.8.1**
>
> Let (X, d) be a dislocated metric space, and A, B, S, T be four self maps of X. Suppose the following conditions hold
>
> (a) $A^r(X) \subseteq S^r(X)$, and $B^r(X) \subseteq T^r(X)$, for any $r \in \mathbb{N}$
>
> (b) (A, B) is a higher-order PS-contraction of type II with respect to (S, T)
>
> (c) the pairs (A, T) or (B, S) satisfy r-(CLR) property
>
> (d) the pairs (A, T) and (B, S) are weakly r-compatible
>
> Then
>
> (i) The maps A and T have a r-coincidence point
>
> (ii) The maps B and S have a r-coincidence point
>
> (iii) The maps A, B, S, T have a unique common r-fixed point

Prove that if $A = B$ in Theorem 7.8.1, then we get the following

> **Corollary 7.8.2**
>
> Let (X, d) be a dislocated metric space, and A, S, T be three self maps of X. Suppose the following conditions hold
>
> (a) $A^r(X) \subseteq S^r(X)$, and $A^r(X) \subseteq T^r(X)$, for any $r \in \mathbb{N}$
>
> (b) A is a higher-order PS-contraction of type II with respect to (S, T), that is,
>
> $$d(A^r x, A^r y) \leq Z' \beta^r Q(x, y)$$
>
> where
>
> $$Q(x, y) := \max\{d(Sy, Ax), d(Tx, Sy), d(Tx, Ax), \\ d(Ay, Sy), d(Tx, Ay)\}$$
>
> $Z' \geq 1$ is modified from Proposition 7.6.2 and $\beta \in [0, \frac{1}{2})$
>
> (c) the pairs (A, T) and (A, S) satisfy r-(CLR) property
>
> (d) the pairs (A, T) and (A, S) are weakly r-compatible
>
> Then
>
> (i) The maps A and T have a r-coincidence point
>
> (ii) The maps A and S have a r-coincidence point
>
> (iii) The maps A, S, T have a unique common r-fixed point

Prove that if $S = T$ in Theorem 7.8.1, then we get the following

> **Corollary 7.8.3**
>
> Let (X, d) be a dislocated metric space, and A, B, S be three self maps of X. Suppose the following conditions hold
>
> (a) $A^r(X) \subseteq S^r(X)$, and $B^r(X) \subseteq S^r(X)$, for any $r \in \mathbb{N}$
>
> (b) (A, B) is a higher-order PS-contraction of type II with respect to S, that is,
> $$d(A^r x, B^r y) \leq Z'' \beta^r R(x, y)$$
> where
> $$R(x, y) := \max\{d(Sy, Ax), d(Sx, Sy), d(Sx, Ax),$$
> $$d(By, Sy), d(Sx, By)\}$$
> $Z'' \geq 1$ is modified from Proposition 7.6.2 and $\beta \in [0, \frac{1}{2})$
>
> (c) the pairs (A, S) or (B, S) satisfy r-(CLR) property
>
> (d) the pairs (A, S) and (B, S) are weakly r-compatible
>
> Then
>
> (i) The maps A and S have a r-coincidence point
>
> (ii) The maps B and S have a r-coincidence point
>
> (iii) The maps A, B, S have a unique common r-fixed point

Prove that if $A = B$ and $S = T$ in Theorem 7.8.1, then we get the following

> **Corollary 7.8.4**
>
> Let (X, d) be a dislocated metric space, and A, S be two self maps of X. Suppose the following conditions hold
>
> (a) $A^r(X) \subseteq S^r(X)$, for any $r \in \mathbb{N}$
>
> (b) A is a higher-order PS-contraction of type II with respect to S, that is,
> $$d(A^r x, A^r y) \leq Z''' \beta^r K(x, y)$$
> where
> $$K(x, y) := \max\{d(Sy, Ax), d(Sx, Sy), d(Sx, Ax),$$
> $$d(Ay, Sy), d(Sx, Ay)\}$$
> $Z''' \geq 1$ is modified from Proposition 7.6.2 and $\beta \in [0, \frac{1}{2})$
>
> (c) the pair (A, S) satisfy r-(CLR) property
>
> (d) the pair (A, S) is weakly r-compatible
>
> Then
>
> (i) The maps A and S have a r-coincidence point
>
> (ii) The maps A and S have a unique r-common fixed point

Chapter 8

Common r-Fixed Point Theorems in b-Dislocated Metric Space using r-(E.A) and r-(E.A)-Like Properties

8.1 Brief Summary

> **Abstract**
>
> Inspired by higher-order fixed point theory [Clement Ampadu, Fixed Point Theory for Higher-Order Mappings. ISBN: 5800118959925, lulu.com, 2016], and the concepts of $(E.A)$ property and $(E.A)$-like property, see for example, [M. Aamri and D. El Moutawakil, Some new common fixed point theorems under strict contractive conditions, J. Math. Anal. Appl. 270, 181-188, 2002; K. Wadhwa, H. Dubey and R. Jain, Impact of E. A. Like property on common fixed point theorems in fuzzy metric spaces. J. Adv. Stud. Topology 3 (1) (2012), 52-59] we obtain some higher-order common fixed point theorems in the setting of b-dislocated metric spaces which derive from certain results contained in [Kastriot Zoto, Ilir Vardhami, Jani Dine and Arben Isufati, COMMON FIXED POINTS IN b-DISLOCATED METRIC SPACES USING (E.A) PROPERTY, BULLETIN MATHÉMATIQUE, Vol. 40(LXVI) No. 1 2014 (15-27)]

8.2 Introduction and Preliminaries

> **Definition 8.2.1**
>
> [R. Shrivastava, Z. K. Ansari and M. Sharma, Some results on Fixed Points in Dislocated and Dislocated Quasi-Metric Spaces, Journal of Advanced Studies in Topology, Vol. 3, No.1, (2012)] Let X be a nonempty set. A mapping $d_l : X \times X \mapsto [0, \infty)$ is called a dislocated metric (or simply d_l-metric) if the following conditions hold for all $x, y, z \in X$
>
> (a) If $d_l(x, y) = 0$, then $x = y$
>
> (b) $d_l(x, y) = d_l(y, x)$
>
> (c) $d_l(x, y) \leq d_l(x, z) + d_l(z, y)$
>
> and we say (X, d_l) is a dislocated metric space.

Remark 8.2.2

If $x = y$, then $d_l(x,y)$ may not be zero

Definition 8.2.3

[M. A. Kutbi, M. Arshad, J. Ahmad, A. Azam, Generalized common fixed point results with applications, Abstract and Applied Analysis, volume 2014, article ID 363925, 7 pages] Let X be a nonempty set. A mapping $b_d : X \times X \mapsto [0,\infty)$ is called a b-dislocated metric if the following conditions hold for all $x,y,z \in X$ and $s \geq 1$

(a) If $b_d(x,y) = 0$, then $x = y$

(b) $b_d(x,y) = b_d(y,x)$

(c) $b_d(x,y) \leq s[b_d(x,z) + b_d(z,y)]$

In particular the pair (X, b_d) is called a b-dislocated metric space

Remark 8.2.4

A b-dislocated metric space is a dislocated metric space when $s = 1$, thus the class of b-dislocated metric space is larger than that of dislocated metric.

Example 8.2.5

[Kastriot Zoto, Ilir Vardhami, Jani Dine and Arben Isufati, COMMON FIXED POINTS IN b-DISLOCATED METRIC SPACES USING (E.A) PROPERTY, BULLETIN MATHÉMATIQUE, Vol. 40(LXVI) No. 1 2014 (15-27)] Let $X = \mathbb{R}$, then $d_l(x,y) = |x| + |y|$ defines a dislocated metric on X

Example 8.2.6

[Kastriot Zoto, Ilir Vardhami, Jani Dine and Arben Isufati, COMMON FIXED POINTS IN b-DISLOCATED METRIC SPACES USING (E.A) PROPERTY, BULLETIN MATHÉMATIQUE, Vol. 40(LXVI) No. 1 2014 (15-27)] Let $X = \mathbb{R}^+ \cup \{0\}$, then $b_d(x,y) = (x+y)^2$ defines a b-dislocated metric on X with parameter $s = 2$

Definition 8.2.7

[N. Hussain, J. R. Roshan, V. Parvaneh and M. Abbas, Common fixed point results for weak contractive mappings in ordered b-dislocated metric spaces with applications, Journal of inequalities and Applications, 1/486, (2013)] Let (X, b_d) be a b_d-metric space, and $\{x_n\}$ be a sequence of points in X. A point $x \in X$ is said to be the limit of the sequence $\{x_n\}$ if $\lim_{n \to \infty} b_d(x_n, x) = 0$, and we say that the sequence $\{x_n\}$ is b_d-convergent to x and denote it by $x_n \to x$ as $n \to \infty$

Proposition 8.2.8

[M. A. Kutbi, M. Arshad, J. Ahmad, A. Azam, Generalized common fixed point results with applications, Abstract and Applied Analysis, volume 2014, article ID 363925, 7 pages] The limit of a b_d-convergent sequence in a b_d-metric space is unique

Definition 8.2.9

[N. Hussain, J. R. Roshan, V. Parvaneh and M. Abbas, Common fixed point results for weak contractive mappings in ordered b-dislocated metric spaces with applications, Journal of inequalities and Applications, 1/486, (2013)] A sequence $\{x_n\}$ in a b_d-metric space (X, b_d) is called a b_d-Cauchy sequence iff given $\epsilon > 0$, there is $n_0 \in \mathbb{N}$ such that for all $n, m > n_0$, we have, $b_d(x_n, x_m) < \epsilon$ or $\lim_{n,m \to \infty} b_d(x_n, x_m) = 0$

Remark 8.2.10

[Kastriot Zoto, Ilir Vardhami, Jani Dine and Arben Isufati, COMMON FIXED POINTS IN b-DISLOCATED METRIC SPACES USING (E.A) PROPERTY, BULLETIN MATHÉMATIQUE, Vol. 40(LXVI) No. 1 2014 (15-27)] Every b_d-convergent sequence in a b_d-metric space is a b_d-Cauchy sequence

Definition 8.2.11

[N. Hussain, J. R. Roshan, V. Parvaneh and M. Abbas, Common fixed point results for weak contractive mappings in ordered b-dislocated metric spaces with applications, Journal of inequalities and Applications, 1/486, (2013)] A b_d-metric space (X, b_d) is called complete if every b_d Cauchy sequence in X is b_d-convergent in X

Taking inspiration from [I. Beg and M. Abbas, Coincidence and common fixed points of noncompatible maps. J. Appl. Math. Inform. 29(3-4), 743-752 (2011)], we introduce the following

Definition 8.2.12

Let f and g be two self-maps on a metric space (X, d). We say f and g are r-compatible if for any $r \in \mathbb{N}$

$$\lim_{n \to \infty} d(f^r g^r x_n, g^r f^r x_n) = \lim_{n \to \infty} d(f^r x_n, g^r x_n) = 0$$

whenever $\{x_n\}$ is a sequence in X such that $\lim_{n \to \infty} f^r x_n = \lim_{n \to \infty} g^r x_n = z$ for some $z \in X$

Taking inspiration from [G. Jungck, Compatible mappings and common fixed points, Int. J. Math. Math. Sci., (1986), 771-779] we introduce the following

Definition 8.2.13

Let f and g be self mappings of a set X. We say f and g are weakly r-compatible if they r-commute at their r-coincidence point, that is, if for any $r \in \mathbb{N}$, $f^r x = g^r x$ for some $x \in X$ implies $g^r f^r x = f^r g^r x$

Lemma 8.2.14

[N. Hussain, J. R. Roshan, V. Parvaneh and M. Abbas, Common fixed point results for weak contractive mappings in ordered b-dislocated metric spaces with applications, Journal of inequalities and Applications, 1/486, (2013)] Let (X, b_d) be a b-dislocated metric space with parameter $s \geq 1$. Suppose that $\{x_n\}$ and $\{y_n\}$ are b_d-convergent to $x, y \in X$, respectively. Then we have

$$\frac{1}{s^2} b_d(x, y) \leq \liminf_{n \to \infty} b_d(x_n, y_n) \leq \limsup_{n \to \infty} b_d(x_n, y_n) \leq s^2 b_d(x, y)$$

In particular, if $b_d(x, y) = 0$, then we have $\lim_{n \to \infty} b_d(x_n, y_n) = 0 = b_d(x, y)$. Moreover for each $z \in X$, we have,

$$\frac{1}{s} b_d(x, z) \leq \liminf_{n \to \infty} b_d(x_n, z) \leq \limsup_{n \to \infty} b_d(x_n, z) \leq s b_d(x, z)$$

In particular, if $b_d(x, z) = 0$, then we have $\lim_{n \to \infty} b_d(x_n, z) = 0 = b_d(x, z)$.

The next two definitions are inspired by [M. Aamri and D. El Moutawakil Some new common fixed point theorems under strict contractive conditions, J. Math. Anal. Appl. 270, 181-188, 2002; K. Wadhwa, H. Dubey and R. Jain, Impact of E. A. Like property on common fixed point theorems in fuzzy metric spaces. J. Adv. Stud. Topology 3 (1) (2012), 52-59].

> **Definition 8.2.15**
>
> Let X be a b-dislocated metric space. Selfmaps f and g on X are said to satisfy r-(E.A) property if there exists a sequence $\{x_n\} \in X$ such that for any $r \in \mathbb{N}$
>
> $$\lim_{n \to \infty} b_d(f^r x_n, t) = \lim_{n \to \infty} b_d(g^r x_n, t) = b_d(t, t) = 0$$
>
> for some $t \in X$

> **Definition 8.2.16**
>
> Let X be a b-dislocated metric space. Selfmaps f and g on X are said to satisfy r-(E.A)-like property if there exists a sequence $\{x_n\} \in X$ such that for any $r \in \mathbb{N}$
>
> $$\lim_{n \to \infty} b_d(f^r x_n, t) = \lim_{n \to \infty} b_d(g^r x_n, t) = b_d(t, t) = 0$$
>
> for some $t \in f^r(X) \cup g^r(X)$

Taking inspiration from [R. Shrivastava, Z. K. Ansari and M. Sharma, Some results on Fixed Points in Dislocated and Dislocated Quasi-Metric Spaces, Journal of Advanced Studies in Topology, Vol. 3, No.1, (2012)] we introduce the following

> **Definition 8.2.17**
>
> Let f and g be two self-mappings on a nonempty set X then,
>
> (a) For any $r \in \mathbb{N}$, we say $x \in X$ is an r-fixed point of f if $f^r x = x$
>
> (b) For any $r \in \mathbb{N}$, we say x is an r-coincidence point of f and g if $f^r x = g^r x$. Moreover, $u = f^r x = g^r x$ is a point of r-coincidence of f and g
>
> (c) For any $r \in \mathbb{N}$, we say x is a common r-fixed point of f and g, if $f^r x = g^r x = x$

8.3 Main Results

> **Notation 8.3.1**
>
> Ψ will denote the set of all continuous and nondecreasing functions $\psi : [0, \infty) \mapsto [0, \infty)$ such that $\psi(t) = 0$ ifff $t = 0$

Inspired by the contractive inequality in Theorem 3.1 [Kastriot Zoto, Ilir Vardhami, Jani Dine and Arben Isufati, COMMON FIXED POINTS IN b-DISLOCATED METRIC SPACES USING (E.A) PROPERTY, BULLETIN MATHÉMATIQUE, Vol. 40(LXVI) No. 1 2014 (15-27)] we introduce the following

> **Definition 8.3.2**
>
> Let (X, b_d) be a b-dislocated metric space with parameter $s \geq 1$, and let $f, g : X \mapsto X$ be two self-maps of X. We say f is a higher-order ψ-contraction with respect to g if the following holds for all $x, y \in X$
>
> $$\psi(2s^2 b_d(f^r x, f^r y)) \leq \sum_{q=0}^{r-1} \Big\{ c_q \psi \Big(\max \Big\{ b_d(g^{q+1} x, g^{q+1} y), b_d(f^{q+1} x, g^{q+1} x),$$
> $$b_d(f^{q+1} y, g^{q+1} y), \frac{b_d(f^{q+1} x, g^{q+1} y) + b_d(f^{q+1} y, g^{q+1} x)}{2s} \Big\} \Big) \Big\}$$
>
> where $0 \leq c_q < 1$ for all $0 \leq q \leq r-1$ and $r \in \mathbb{N}$

CHAPTER 8. COMMON R-FIXED POINT THEOREMS IN B-DISLOCATED METRIC SPACE USING R-(E.A) AND R-(E.A)-LIKE PROPERTIES

> **Proposition 8.3.3**
>
> Let f,g be two self maps of a b-dislocated metric space (X, b_d) with parameter $s \geq 1$, where f is a higher-order ψ-contraction with respect to g. Put
>
> $$M(x,y) := \psi\left(\max \left\{ b_d(gx, gy), b_d(fx, gx), \right.\right.$$
> $$\left.\left. b_d(fy, gy), \frac{b_d(fx, gy) + b_d(fy, gx)}{2s} \right\}\right)$$
>
> Now for every pair $x \neq y$, define
>
> $$Z := Z(x,y) = \max_{0 \leq v \leq r-1} \beta^{-v} \frac{\psi(2s^2 b_d(f^v x, f^v y))}{M(x,y)}$$
>
> then
>
> $$Z = \max_{n \in \mathbb{N} \cup \{0\}} \beta^{-n} \frac{\psi(2s^2 b_d(f^n x, f^n y))}{M(x,y)}$$
>
> where $\beta \in [0,1)$

Now from Definition 8.3.2 and Proposition 8.3.3 we have

> **Definition 8.3.4**
>
> Let f,g be two self maps of a b-dislocated metric space (X, b_d) with parameter $s \geq 1$. Put
>
> $$M(x,y) := \psi\left(\max \left\{ b_d(gx, gy), b_d(fx, gx), \right.\right.$$
> $$\left.\left. b_d(fy, gy), \frac{b_d(fx, gy) + b_d(fy, gx)}{2s} \right\}\right)$$
>
> We say f is a higher-order ψ-contraction with respect to g if the following holds for all $x, y \in X$ and any $r \in \mathbb{N}$
>
> $$\psi(2s^2 b_d(f^r x, f^r y)) \leq Z \beta^r M(x,y)$$
>
> where $Z \geq 1$ is given by the previous Propositon and $\beta \in [0,1)$

Now we prove the following

> **Theorem 8.3.5**
>
> Let (X, b_d) be a b-dislocated metric space with parameter $s \geq 1$, and $f, g : X \mapsto X$ be two self mappings where f is a higher-order ψ-contraction with respect to g. Suppose the pair (f,g) satisfy r-(E.A)-like property in X, then the pair (f,g) have a common point of r-coincidence in X. Moreover, if the pair (f,g) is weakly r-compatible, then f, g have a unique common r-fixed point in X

> **Proof of Theorem 8.3.5**
>
> Since f and g satisfy the r-(E.A)-like property, there exists a sequence $\{x_n\} \in X$ such that $\lim_{n\to\infty} f^r x_n = \lim_{n\to\infty} g^r x_n = t$ for some $t \in f^r(X) \cup g^r(X)$. Assuming that $\lim_{n\to\infty} f^r x_n = t \in g^r(X)$, then, $t = g^r u$ for some $u \in X$. Now since f is a higher-order ψ-contraction with respect to g, we have the following
>
> $$\psi(2s^2 b_d(f^r u, f^r x_n)) \leq Z\beta^r M(u, x_n)$$
>
> where
>
> $$M(u, x_n) := \psi\left(\max\left\{b_d(g^r u, g^r x_n), b_d(f^r u, g^r u),\right.\right.$$
> $$\left.\left. b_d(f^r x_n, g^r x_n), \frac{b_d(f^r u, g^r x_n) + b_d(f^r x_n, g^r u)}{2s}\right\}\right)$$
>
> Since f and g satisfy the r-(E.A)-like property, then using Lemma 8.14 and taking upper limit in the above as $n \to \infty$, we deduce the following
>
> $$\psi(2s b_d(f^r u, t)) = \psi(2s^2 \frac{1}{s} b_d(f^r u, t)) \leq Z\beta^r M(u, t)$$
>
> where $M(u,t) = \psi(\max\{0, b_d(f^r u, t), 0, \frac{b_d(f^r u, t)}{2}\})$. It now follows that
>
> $$\psi(2s b_d(f^r u, t)) \leq Z\beta^r \psi(b_d(f^r u, t))$$
>
> By the properties of ψ, since $Z\beta^r < 1$ and $s \geq 1$ the above inequality implies $\psi(b_d(f^r u, t)) = 0$, that is, $f^r u = t$. It now follows that u is an r-coincidence point of f and g, that is, $f^r u = g^r u = t$. Since f and g are weakly r-compatible, it follows that
>
> $$f^r t = f^r g^r u = g^r f^r u = g^r t$$
>
> Now we show that t is an r-fixed point of f. Now since f is a higher-order ψ-contraction with respect to g, we have the following
>
> $$\psi(2s^2 b_d(f^r t, f^r x_n)) \leq Z\beta^r M(t, x_n)$$
>
> where
>
> $$M(t, x_n) := \psi\left(\max\left\{b_d(g^r t, g^r x_n), b_d(f^r t, g^r t),\right.\right.$$
> $$\left.\left. b_d(f^r x_n, g^r x_n), \frac{b_d(f^r t, g^r x_n) + b_d(f^r x_n, g^r t)}{2s}\right\}\right)$$
>
> Since f and g satisfy the r-(E.A)-like property, then using Lemma 8.14 and taking upper limit in the above as $n \to \infty$, we deduce the following
>
> $$\psi(2s b_d(f^r t, t)) = \psi(2s^2 \frac{1}{s} b_d(f^r t, t))$$
> $$\leq Z\beta^r \psi\left(\max\left\{s b_d(f^r t, t), b_d(f^r t, f^r t), 0, \frac{s b_d(f^r t, t) + 0}{2s}\right\}\right)$$
> $$\leq Z\beta^r \psi(2s b_d(f^r t, t))$$
>
> The above inequality implies $\psi(2s b_d(f^r t, t)) = 0$. Consequently, we deduce that $f^r t = g^r t = t$. Hence t is a common r-fixed point of f and g.

> **Proof of Theorem 8.3.5 Continued**
>
> For uniqueness let $t \neq t_1$ be two common r-fixed points of the mappings f and g. Now observe we have the following
>
> $$\psi(2sb_d(f^r t, f^r t_1)) \leq \psi(2s^2 b_d(f^r t, f^r t_1)) \leq Z\beta^r M(t, t_1)$$
>
> where
>
> $$M(t, t_1) := \psi\left(\max\left\{b_d(g^r t, g^r t_1), b_d(f^r t, g^r t),\right.\right.$$
> $$\left.\left. b_d(f^r t_1, g^r t_1), \frac{b_d(f^r t, g^r t_1) + b_d(f^r t_1, g^r t)}{2s}\right\}\right)$$
> $$= \psi\left(\max\left\{b_d(t, t_1), b_d(t, t),\right.\right.$$
> $$\left.\left. b_d(t_1, t_1), \frac{b_d(t, t_1) + b_d(t_1, t)}{2s}\right\}\right)$$
> $$\leq \psi(2s b_d(t, t_1))$$
>
> Thus, we deduce that
>
> $$\psi(2s b_d(t, t_1)) = \psi(2s b_d(f^r t, f^r t_1)) \leq Z\beta^r \psi(2s b_d(t, t_1))$$
>
> From the above it is clear that $\psi(2s b_d(t, t_1)) = 0$, and by the property of ψ, we deduce that $2s b_d(t, t_1) = 0$, and hence $t = t_1$. Thus, uniqueness follows, and the proof is finished.

Let Z' be the modification from Proposition 8.3.3 when $\psi(t) = t$, then we get the following

> **Corollary 8.3.6**
>
> Let (X, b_d) be a b-dislocated metric space with parameter $s \geq 1$, and $f, g : X \mapsto X$ be two self mappings where f is a higher-order contraction with respect to g, in the sense that, the following holds for all $x, y \in X$ and any $r \in \mathbb{N}$
>
> $$2s^2 b_d(f^r x, f^r y) \leq Z' \beta^r Q(x, y)$$
>
> where
>
> $$Q(x, y) := \max\left\{b_d(gx, gy), b_d(fx, gx),\right.$$
> $$\left. b_d(fy, gy), \frac{b_d(fx, gy) + b_d(fy, gx)}{2s}\right\}$$
>
> Z' is modified from Proposition 8.3.3, and $\beta \in [0, 1)$. Suppose the pair (f, g) satisfy r-(E.A)-like property in X, then the pair (f, g) have a common point of r-coincidence in X. Moreover, if the pair (f, g) is weakly r-compatible, then f, g have a unique common r-fixed point in X

> **Theorem 8.3.7**
>
> Let (X, b_d) be a complete b-dislocated metric space with parameter $s \geq 1$, and $f, g : X \mapsto X$ be two self mappings where f is a higher-order ψ-contraction with respect to g, and for any $r \in \mathbb{N}$, $f^r(X) \subseteq g^r(X)$. Suppose the pair (f, g) satisfy r-(E.A) property in X and $g^r(X)$ is b_d-closed in X for any $r \in \mathbb{N}$, then the pair (f, g) have a common point of r-coincidence in X. Moreover, if the pair (f, g) is weakly r-compatible, then f, g have a unique common r-fixed point in X

> **Proof of Theorem 8.3.7**
>
> Since f and g satisfy r-(E.A) property, there exists a sequence $\{x_n\}$ in X such that $\lim_{n\to\infty} f^r x_n = \lim_{n\to\infty} g^r x_n = t$ for some $t \in X$. As $g^r(X)$ is a b_d-closed subspace of X for any $r \in \mathbb{N}$, it follows that every convergent sequence of points of $g^r(X)$ has a limit in $g^r(X)$. Thus,
> $$t = \lim_{n\to\infty} f^r x_n = \lim_{n\to\infty} g^r x_n = g^r u$$
> for some $u \in X$. It follows that $t = g^r(u) \in g^r(X)$, and in this condition the pair (f,g) satisfy r-(E.A)-like property, and the proof follows from Theorem 8.3.5

8.4 Open Problem I

> **Definition 8.4.1**
>
> Let (X, b_d) be a b-dislocated metric space with parameter $s \geq 1$, and $f, g : X \mapsto X$ be two self mappings. We say f is a higher-order $(\alpha, \beta, \gamma, \delta)$-contraction with respect to g if the following holds for all $x, y \in X$
>
> $$s^2 b_d(f^r x, f^r y)) \leq \sum_{q=0}^{r-1} c_q \Big\{ b_d(g^{q+1}x, f^{q+1}y) + b_d(g^{q+1}x, g^{q+1}y) +$$
> $$b_d(g^{q+1}y, f^{q+1}y) + b_d(g^{q+1}x, f^{q+1}x) \Big\}$$
>
> where $0 \leq c_q < \frac{1}{8}$ for all $0 \leq q \leq r-1$ and $r \in \mathbb{N}$

> **Proposition 8.4.2**
>
> Let f, g be two self maps of a b-dislocated metric space (X, b_d) with parameter $s \geq 1$, where f is a higher-order $(\alpha, \beta, \gamma, \delta)$-contraction with respect to g. Put
>
> $$M(x,y) := b_d(gx, fy) + b_d(gx, gy) +$$
> $$b_d(gy, fy) + b_d(gx, fx)$$
>
> Now for every pair $x \neq y$, define
>
> $$Z := Z(x,y) = \max_{0 \leq v \leq r-1} \beta^{-v} \frac{\psi(s^2 b_d(f^v x, f^v y))}{M(x,y)}$$
>
> then
>
> $$Z = \max_{n \in \mathbb{N} \cup \{0\}} \beta^{-n} \frac{\psi(s^2 b_d(f^n x, f^n y))}{M(x,y)}$$
>
> where $\beta \in [0, \frac{1}{8})$

Now combining Definition 8.4.1 and Proposition 8.4.2, we introduce the following

> **Definition 8.4.3**
>
> Let f, g be two self maps of a b-dislocated metric space (X, b_d) with parameter $s \geq 1$. We say f is a higher-order $(\alpha, \beta, \gamma, \delta)$-contraction with respect to g if the following holds for all $x, y \in X$ and $r \in \mathbb{N}$
> $$s^2 b_d(f^r x, f^r y)) \leq Z \beta^r M(x,y)$$
> where $Z \geq 1$ is given by the previous Proposition, $\beta \in [0, \frac{1}{8})$, and
> $$M(x,y) := b_d(gx, fy) + b_d(gx, gy) +$$
> $$b_d(gy, fy) + b_d(gx, fx)$$

Now the open problem is to prove the following

> **Theorem 8.4.4**
>
> Let f, g be two self maps of a b-dislocated metric space (X, b_d) with parameter $s \geq 1$, where f is a higher-order $(\alpha, \beta, \gamma, \delta)$-contraction with respect to g. Suppose that the pair (f, g) satisfy r-(E.A)-like property in X. Then the pair (f, g) have a common point of r-coincidence in X. Moreover if the pair (f, g) is weakly r-compatible, then f and g have a unique common r-fixed point in X.

8.5 Open Problem II

In relation to Open Problem I, the open problem here is to prove the following

> **Theorem 8.5.1**
>
> Let f, g be two self maps of a b-dislocated metric space (X, b_d) with parameter $s \geq 1$, where f is a higher-order $(\alpha, \beta, \gamma, \delta)$-contraction with respect to g, and $f^r(X) \subseteq g^r(X)$, for any $r \in \mathbb{N}$. Suppose that the pair (f, g) satisfy r-(E.A) property and $g^r(X)$ is b_d-closed in X for any $r \in \mathbb{N}$. Then the pair (f, g) have a common point of r-coincidence in X. Moreover if the pair (f, g) is weakly r-compatible, then f and g have a unique common r-fixed point in X.

Chapter 9

Common r-Fixed Point Theorems for Mappings in Dislocated Metric Space under r-Compatibility of Type (A)

9.1 Brief Summary

> **Abstract**
>
> Inspired by higher-order fixed point theory [Clement Ampadu, Fixed Point Theory for Higher-Order Mappings. ISBN: 5800118959925, lulu.com, 2016], we obtain the higher-order version of Theorem 2 [Dinesh Panthi and P. Sumati Kumari, Common Fixed Point Theorems for Mappings of Compatible Type(A) in Dislocated Metric Space, Nepal Journal of Science and Technology Vol. 16, No.1 (2015) 79-86], and deduce several Corollaries as a consequence

9.2 Introduction and Preliminaries

> **Definition 9.2.1**
>
> [Dinesh Panthi and P. Sumati Kumari, Common Fixed Point Theorems for Mappings of Compatible Type(A) in Dislocated Metric Space, Nepal Journal of Science and Technology Vol. 16, No.1 (2015) 79-86] Let X be a nonempty set and let $d : X \times X \mapsto [0, \infty)$ be a function satisfying the following conditions for all $x, y, z \in X$
>
> (a) $d(x, y) = d(y, x)$
>
> (b) $d(x, y) = d(y, x) = 0$ implies $x = y$
>
> (c) $d(x, y) \leq d(x, z) + d(z, y)$
>
> Then d is called a dislocated metric (or simply d-metric) on X

> **Definition 9.2.2**
>
> [Dinesh Panthi and P. Sumati Kumari, Common Fixed Point Theorems for Mappings of Compatible Type(A) in Dislocated Metric Space, Nepal Journal of Science and Technology Vol. 16, No.1 (2015) 79-86] A sequence $\{x_n\}$ in a d-metric space (X, d) is called a Cauchy sequence if for a given $\epsilon > 0$, there corresponds $n_0 \in \mathbb{N}$ such that for all $m, n \geq n_0$, we have $d(x_n, x_m) < \epsilon$

Definition 9.2.3

[Dinesh Panthi and P. Sumati Kumari, Common Fixed Point Theorems for Mappings of Compatible Type(A) in Dislocated Metric Space, Nepal Journal of Science and Technology Vol. 16, No.1 (2015) 79-86] A sequence in a d-metric space converges with respect to d if there exists $x \in X$ such that $\lim_{n \to \infty} d(x_n, x) = 0$. In this case, x is called the limit of $\{x_n\}$ (in d), and we write $x_n \to x$

Definition 9.2.4

[Dinesh Panthi and P. Sumati Kumari, Common Fixed Point Theorems for Mappings of Compatible Type(A) in Dislocated Metric Space, Nepal Journal of Science and Technology Vol. 16, No.1 (2015) 79-86] A d-metric space (X, d) is called complete if every Cauchy sequence in X converges with respect to d to an element in X

Lemma 9.2.5

[Dinesh Panthi and P. Sumati Kumari, Common Fixed Point Theorems for Mappings of Compatible Type(A) in Dislocated Metric Space, Nepal Journal of Science and Technology Vol. 16, No.1 (2015) 79-86] Limits in a d-metric space are unique

The concept of commuting mappings have appeared in the literature, and for example see [Dinesh Panthi and P. Sumati Kumari, Common Fixed Point Theorems for Mappings of Compatible Type(A) in Dislocated Metric Space, Nepal Journal of Science and Technology Vol. 16, No.1 (2015) 79-86], now we introduce the following

Definition 9.2.6

Let A and S be two self mappings on a set X. We say mappings A and S are r-commuting if
$$A^r S^r x = S^r A^r x$$
for all $x \in X$ and any $r \in \mathbb{N}$

The concept of point of coincidence of two self mappings have appeared in the literature, and for example see [Dinesh Panthi and P. Sumati Kumari, Common Fixed Point Theorems for Mappings of Compatible Type(A) in Dislocated Metric Space, Nepal Journal of Science and Technology Vol. 16, No.1 (2015) 79-86], now we introduce the following

Definition 9.2.7

Let A and S be two self mappings on a set X. If $A^r x = S^r x$ for some $x \in X$ and any $r \in \mathbb{N}$, then we say x is a r-coincidence point of A and S

The notion of compatible mappings of type (A) have appeared in the literature, and for example see [Dinesh Panthi and P. Sumati Kumari, Common Fixed Point Theorems for Mappings of Compatible Type(A) in Dislocated Metric Space, Nepal Journal of Science and Technology Vol. 16, No.1 (2015) 79-86], now we introduce the following

Definition 9.2.8

Two mappings S and T from a metric space (X, d) into itself will be called r-compatible of type (A) if
$$\lim_{n \to \infty} d(S^r T^r x_n, T^r T^r x_n) = 0$$
and
$$\lim_{n \to \infty} d(T^r S^r x_n, S^r S^r x_n) = 0$$
whenever $\{x_n\}$ is a sequence in X such that $\lim_{n \to \infty} S^r x_n = \lim_{n \to \infty} T^r x_n = x$ for some $x \in X$ and any $r \in \mathbb{N}$

> **Proposition 9.2.9**
>
> Let S and T be mappings from a metric space (X,d) into itself which are r-compatible of type (A). Suppose that $\lim_{n\to\infty} S^r x_n = \lim_{n\to\infty} T^r x_n = x$ for some $x \in X$ and any $r \in \mathbb{N}$. If S is r-continuous, that is, S^r is continuous for any $r \in \mathbb{N}$, then $\lim_{n\to\infty} T^r S^r x_n = S^r x$ for any $r \in \mathbb{N}$

> **Proof of Proposition 9.2.9**
>
> Since S is r-continuous, it follows that
>
> $$S^r x_n \to x \implies S^r S^r x_n \to S^r x$$
>
> and
>
> $$T^r x_n \to x \implies S^r T^r x_n \to S^r x$$
>
> Now observe that
>
> $$d(T^r S^r x_n, S^r x) \leq d(T^r S^r x_n, S^r S^r x_n) + d(S^r S^r x_n, S^r T^r x_n) + d(S^r T^r x_n, S^r x)$$
>
> Now taking limits in the above as $n \to \infty$, we get $\lim_{n\to\infty} d(T^r S^r x_n, S^r x) = 0$ which implies $\lim_{n\to\infty} T^r S^r x_n = S^r x$

9.3 Main Results

Inspired by the contractive condition in Theorem 2 [Dinesh Panthi and P. Sumati Kumari, Common Fixed Point Theorems for Mappings of Compatible Type(A) in Dislocated Metric Space, Nepal Journal of Science and Technology Vol. 16, No.1 (2015) 79-86], we introduce the following

> **Definition 9.3.1**
>
> Let (X,d) be a d-metric space, and $A, B, S, T : X \mapsto X$ be four self mappings. We say (T,S) is a $(\alpha, \beta, \gamma, \kappa, \delta, \mu)$- type contraction with respect to (A,B) if the following holds for all $x, y \in X$ and $k \in [0, \frac{1}{6})$
>
> $$d(Tx, Sy) \leq kM(x,y)$$
>
> where
>
> $$M(x,y) = \frac{d(Ay, Sy)d(Bx, Ay)}{d(Ax, Tx) + d(Sy, Ax)} + \frac{d(Tx, Ax)d(Ty, By)}{d(Ax, Tx) + d(Sy, Ax)}$$
> $$+ \frac{d(Ax, Sx)d(Sy, Ay)}{d(Ax, Tx) + d(Sy, Ax)} + \frac{d(Bx, Ay)d(Tx, Sy)}{d(Bx, Tx) + d(Bx, Sy)}$$
> $$+ \frac{d(Bx, Tx)d(Ay, Sy)}{d(Ax, Ty) + d(Bx, Ty)} + \frac{d(Ax, Sx)d(By, Ty)}{d(Bx, Ay) + d(Ax, Ty)}$$

Now the higher-order version of the above reads as follows

CHAPTER 9. COMMON R-FIXED POINT THEOREMS FOR MAPPINGS IN DISLOCATED METRIC SPACE UNDER R-COMPATIBILITY OF TYPE (A)

Definition 9.3.2

Let (X, d) be a d-metric space, and $A, B, S, T : X \mapsto X$ be four self mappings. We say (T, S) is a higher-order $(\alpha, \beta, \gamma, \kappa, \delta, \mu)$- type contraction with respect to (A, B) if the following holds for all $x, y \in X$, $0 \leq c_q < \frac{1}{6}$, $0 \leq q \leq r-1$, and $r \in \mathbb{N}$

$$d(T^r x, S^r y) \leq \sum_{q=0}^{r-1} c_q V(x, y)$$

where

$$V(x,y) = \frac{d(A^{q+1}y, S^{q+1}y)d(B^{q+1}x, A^{q+1}y)}{d(A^{q+1}x, T^{q+1}x) + d(S^{q+1}y, A^{q+1}x)} + \frac{d(T^{q+1}x, A^{q+1}x)d(T^{q+1}y, B^{q+1}y)}{d(A^{q+1}x, T^{q+1}x) + d(S^{q+1}y, A^{q+1}x)}$$
$$+ \frac{d(A^{q+1}x, S^{q+1}x)d(S^{q+1}y, A^{q+1}y)}{d(A^{q+1}x, T^{q+1}x) + d(S^{q+1}y, A^{q+1}x)} + \frac{d(B^{q+1}x, A^{q+1}y)d(T^{q+1}x, S^{q+1}y)}{d(B^{q+1}x, T^{q+1}x) + d(B^{q+1}x, S^{q+1}y)}$$
$$+ \frac{d(B^{q+1}x, T^{q+1}x)d(A^{q+1}y, S^{q+1}y)}{d(A^{q+1}x, T^{q+1}y) + d(B^{q+1}x, T^{q+1}y)} + \frac{d(A^{q+1}x, S^{q+1}x)d(B^{q+1}y, T^{q+1}y)}{d(B^{q+1}x, A^{q+1}y) + d(A^{q+1}x, T^{q+1}y)}$$

Proposition 9.3.3

Let A, B, S, T be four self maps of a d-metric space (X, d), where (T, S) is a higher-order $(\alpha, \beta, \gamma, \kappa, \delta, \mu)$- type contraction with respect to (A, B). Put

$$M(x,y) = \frac{d(Ay, Sy)d(Bx, Ay)}{d(Ax, Tx) + d(Sy, Ax)} + \frac{d(Tx, Ax)d(Ty, By)}{d(Ax, Tx) + d(Sy, Ax)}$$
$$+ \frac{d(Ax, Sx)d(Sy, Ay)}{d(Ax, Tx) + d(Sy, Ax)} + \frac{d(Bx, Ay)d(Tx, Sy)}{d(Bx, Tx) + d(Bx, Sy)}$$
$$+ \frac{d(Bx, Tx)d(Ay, Sy)}{d(Ax, Ty) + d(Bx, Ty)} + \frac{d(Ax, Sx)d(By, Ty)}{d(Bx, Ay) + d(Ax, Ty)}$$

Now for every pair $x \neq y$, define

$$Z := Z(x,y) = \max_{0 \leq v \leq r-1} \beta^{-v} \frac{d(T^v x, S^v y)}{M(x,y)}$$

then

$$Z = \max_{n \in \mathbb{N} \cup \{0\}} \beta^{-n} \frac{d(T^n x, S^n y)}{M(x,y)}$$

where $\beta \in [0, \frac{1}{6})$

Now we have the following alternate characterization of the higher-order $(\alpha, \beta, \gamma, \kappa, \delta, \mu)$- type contraction

Definition 9.3.4

Let A, B, S, T be four self maps of a d-metric space (X, d), we say (T, S) is a higher-order $(\alpha, \beta, \gamma, \kappa, \delta, \mu)$- type contraction with respect to (A, B) if the following holds for all $x, y \in X$ and $r \in \mathbb{N}$, where $Z \geq 1$ is given by the previous Proposition, and $\beta \in [0, \frac{1}{6})$

$$d(T^r x, S^r y) \leq Z\beta^r M(x, y)$$

where

$$M(x, y) = \frac{d(Ay, Sy)d(Bx, Ay)}{d(Ax, Tx) + d(Sy, Ax)} + \frac{d(Tx, Ax)d(Ty, By)}{d(Ax, Tx) + d(Sy, Ax)}$$
$$+ \frac{d(Ax, Sx)d(Sy, Ay)}{d(Ax, Tx) + d(Sy, Ax)} + \frac{d(Bx, Ay)d(Tx, Sy)}{d(Bx, Tx) + d(Bx, Sy)}$$
$$+ \frac{d(Bx, Tx)d(Ay, Sy)}{d(Ax, Ty) + d(Bx, Ty)} + \frac{d(Ax, Sx)d(By, Ty)}{d(Bx, Ay) + d(Ax, Ty)}$$

Now our main result is as follows

Theorem 9.3.5

Let (X, d) be a complete d-metric space, and A, B, S, T be four self maps of X. Suppose the following conditions hold

(a) For any $r \in \mathbb{N}$, $T^r(X) \subset A^r(X)$ and $S^r(X) \subset B^r(X)$

(b) The pairs (T, B) and (S, A) are r-compatible of type (A)

(c) (T, S) is a higher-order $(\alpha, \beta, \gamma, \kappa, \delta, \mu)$- type contraction with respect to (A, B)

If one of A, B, S, T is r-continuous, then A, B, S, T have a unique common r-fixed point in X.

CHAPTER 9. COMMON R-FIXED POINT THEOREMS FOR MAPPINGS IN DISLOCATED METRIC SPACE UNDER R-COMPATIBILITY OF TYPE (A)

Proof of Theorem 9.3.5

Let $\{y_n\} \in X$ be such that $T^r x_{2n+1} = y_{2n+2}$, $A^r x_{2n} = y_{2n}$, $S^r x_{2n+1} = y_{2n+2}$, and $B^r x_{2n} = y_{2n}$ for $n = 1, 2, 3, \cdots$. Now since (T, S) is a higher-order $(\alpha, \beta, \gamma, \kappa, \delta, \mu)$- type contraction with respect to (A, B) we deduce the following

$$d(T^r x_{2n}, S^r y_{2n+1}) \leq Z\beta^r M(x_{2n}, y_{2n+1})$$

where

$$M(x_{2n}, y_{2n+1}) = \frac{d(A^r y_{2n+1}, S^r y_{2n+1}) d(B^r x_{2n}, A^r y_{2n+1})}{d(A^r x_{2n}, T^r x_{2n}) + d(S^r y_{2n+1}, A^r x_{2n})}$$
$$+ \frac{d(T^r x_{2n}, A^r x_{2n}) d(T^r y_{2n+1}, B^r y_{2n+1})}{d(A^r x_{2n}, T^r x_{2n}) + d(S^r y_{2n+1}, A^r x_{2n})}$$
$$+ \frac{d(A^r x_{2n}, S^r x_{2n}) d(S^r y_{2n+1}, A^r y_{2n+1})}{d(A^r x_{2n}, T^r x_{2n}) + d(S^r y_{2n+1}, A^r x_{2n})}$$
$$+ \frac{d(B^r x_{2n}, A^r y_{2n+1}) d(T^r x_{2n}, S^r y_{2n+1})}{d(B^r x_{2n}, T^r x_{2n}) + d(B^r x_{2n}, S^r y_{2n+1})}$$
$$+ \frac{d(B^r x_{2n}, T^r x_{2n}) d(A^r y_{2n+1}, S^r y_{2n+1})}{d(A^r x_{2n}, T^r y_{2n+1}) + d(B^r x_{2n}, T^r y_{2n+1})}$$
$$+ \frac{d(A^r x_{2n}, S^r x_{2n}) d(B^r y_{2n+1}, T^r y_{2n+1})}{d(B^r x_{2n}, A^r y_{2n+1}) + d(A^r x_{2n}, T^r y_{2n+1})}$$

Now we have

$$d(y_{2n+1}, y_{2n+2}) \leq 6Z\beta^r \frac{d(y_{2n+1}, y_{2n+2}) d(y_{2n}, y_{2n+1})}{d(y_{2n}, y_{2n+1}) + d(y_{2n+2}, y_{2n})}$$
$$\leq 6Z\beta^r \frac{(d(y_{2n+1}, y_{2n}) + d(y_{2n}, y_{2n+2})) d(y_{2n}, y_{2n+1})}{d(y_{2n}, y_{2n+1}) + d(y_{2n+2}, y_{2n})}$$
$$\leq 6Z\beta^r d(y_{2n+1}, y_{2n})$$

Now let $h := 6Z\beta^r < 1$, then $d(y_{2n+1}, y_{2n+2}) < h d(y_{2n+1}, y_{2n})$. By induction, we obtain $d(y_{2n+1}, y_{2n+2}) \leq h^n d(y_1, y_0)$. In particular, we have $d(y_n, y_{n+1}) \leq h^n d(y_1, y_0)$. Now for every integer $q > 0$ we have

$$d(y_{n+q}, y_n) \leq d(y_{n+q}, y_{n+q-1}) + \cdots + d(y_{n+2}, y_{n+1}) + d(y_{n+1}, y_n)$$
$$\leq (h^{n+q-1} + \cdots + h^{n+1} + h^n) d(y_1, y_0)$$
$$= h^n (h^{q-1} + \cdots + h + 1) d(y_1, y_0)$$

From the above, and since $h < 1$, it follows that $h^n \to 0$ as $n \to \infty$, and so $d(y_{n+q}, y_n) \to 0$ as $n \to \infty$ for all $q > 0$. Consequently, the sequence $\{y_n\}$ is Cauchy. Since X is complete, there exists a point $z \in X$ such that $\{y_n\} \to z$. Consequently, the subsequences

$$\{A^r x_{2n}\} = \{B^r x_{2n}\} \to z$$

$$\{S^r x_{2n+1}\} = \{T^r x_{2n+1}\} \to z$$

Proof of Theorem 9.3.5 Continued

Suppose T is r-continuous. Since the pair (T, B) is r-compatible of type (A), then by Proposition 9.2.9, we have
$$\lim_{n \to \infty} T^r T^r x_{2n} = T^r z$$
and
$$\lim_{n \to \infty} B^r T^r x_{2n} = T^r z$$

Now we show that z is the r-fixed point of T, that is, $T^r z = z$ for any $r \in \mathbb{N}$. Now since (T, S) is a higher-order $(\alpha, \beta, \gamma, \kappa, \delta, \mu)$-type contraction with respect to (A, B) we deduce the following
$$d(T^r T^r x_{2n}, S^r y_{2n+1}) \leq Z\beta^r M(T^r x_{2n}, y_{2n+1})$$

where

$$M(T^r x_{2n}, y_{2n+1}) = \frac{d(A^r y_{2n+1}, S^r y_{2n+1}) d(B^r T^r x_{2n}, A^r y_{2n+1})}{d(A^r T^r x_{2n}, T^r T^r x_{2n}) + d(S^r y_{2n+1}, A^r T^r x_{2n})}$$
$$+ \frac{d(T^r T^r x_{2n}, A^r T^r x_{2n}) d(T^r y_{2n+1}, B^r y_{2n+1})}{d(A^r T^r x_{2n}, T^r T^r x_{2n}) + d(S^r y_{2n+1}, A^r T^r x_{2n})}$$
$$+ \frac{d(A^r T^r x_{2n}, S^r T^r x_{2n}) d(S^r y_{2n+1}, A^r y_{2n+1})}{d(A^r T^r x_{2n}, T^r T^r x_{2n}) + d(S^r y_{2n+1}, A^r T^r x_{2n})}$$
$$+ \frac{d(B^r T^r x_{2n}, A^r y_{2n+1}) d(T^r T^r x_{2n}, S^r y_{2n+1})}{d(B^r T^r x_{2n}, T^r T^r x_{2n}) + d(B^r T^r x_{2n}, S^r y_{2n+1})}$$
$$+ \frac{d(B^r T^r x_{2n}, T^r T^r x_{2n}) d(A^r y_{2n+1}, S^r y_{2n+1})}{d(A^r T^r x_{2n}, T^r y_{2n+1}) + d(B^r T^r x_{2n}, T^r y_{2n+1})}$$
$$+ \frac{d(A^r T^r x_{2n}, S^r T^r x_{2n}) d(B^r y_{2n+1}, T^r y_{2n+1})}{d(B^r T^r x_{2n}, A^r y_{2n+1}) + d(A^r T^r x_{2n}, T^r y_{2n+1})}$$

Now since $\lim_{n \to \infty} T^r T^r x_{2n} = T^r z$, $\lim_{n \to \infty} B^r T^r x_{2n} = T^r z$, $\{A^r x_{2n}\} = \{B^r x_{2n}\} \to z$, and $\{S^r x_{2n+1}\} = \{T^r x_{2n+1}\} \to z$. If we take limits in the above inequality, we deduce that

$$d(T^r z, z) \leq Z\beta^r \frac{d(T^r z, z)^2}{d(T^r z, T^r z) + d(T^r z, z)}$$
$$\leq Z\beta^r \frac{d(T^r z, z)^2}{3 d(T^r z, z)}$$
$$= Z\beta^r \frac{d(T^r z, z)}{3}$$

Since $1 - \frac{Z\beta^r}{3} \neq 0$, the above inequality implies $d(T^r z, z) = 0$, that is, $T^r z = z$. Now assume that B is r-continuous. Since the pair (T, B) is r-compatible of type (A), then
$$B^r B^r x_{2n} \to B^r z$$
and
$$T^r B^r x_{2n} \to B^r z$$

Now we show that z is an r-fixed point of B, that is, $B^r z = z$ for any $r \in \mathbb{N}$. Now since (T, S) is a higher-order $(\alpha, \beta, \gamma, \kappa, \delta, \mu)$-type contraction with respect to (A, B) we deduce the following

Proof of Theorem 9.3.5 Continued

$$d(T^r B^r x_{2n}, S^r y_{2n+1}) \leq Z\beta^r M(B^r x_{2n}, y_{2n+1})$$

where

$$M(B^r x_{2n}, y_{2n+1}) = \frac{d(A^r y_{2n+1}, S^r y_{2n+1}) d(B^r B^r x_{2n}, A^r y_{2n+1})}{d(A^r B^r x_{2n}, T^r B^r x_{2n}) + d(S^r y_{2n+1}, A^r B^r x_{2n})}$$
$$+ \frac{d(T^r B^r x_{2n}, A^r B^r x_{2n}) d(T^r y_{2n+1}, B^r y_{2n+1})}{d(A^r B^r x_{2n}, T^r B^r x_{2n}) + d(S^r y_{2n+1}, A^r B^r x_{2n})}$$
$$+ \frac{d(A^r B^r x_{2n}, S^r B^r x_{2n}) d(S^r y_{2n+1}, A^r y_{2n+1})}{d(A^r B^r x_{2n}, T^r B^r x_{2n}) + d(S^r y_{2n+1}, A^r B^r x_{2n})}$$
$$+ \frac{d(B^r B^r x_{2n}, A^r y_{2n+1}) d(T^r B^r x_{2n}, S^r y_{2n+1})}{d(B^r B^r x_{2n}, T^r B^r x_{2n}) + d(B^r B^r x_{2n}, S^r y_{2n+1})}$$
$$+ \frac{d(B^r B^r x_{2n}, T^r B^r x_{2n}) d(A^r y_{2n+1}, S^r y_{2n+1})}{d(A^r B^r x_{2n}, T^r y_{2n+1}) + d(B^r B^r x_{2n}, T^r y_{2n+1})}$$
$$+ \frac{d(A^r B^r x_{2n}, S^r B^r x_{2n}) d(B^r y_{2n+1}, T^r y_{2n+1})}{d(B^r B^r x_{2n}, A^r y_{2n+1}) + d(A^r B^r x_{2n}, T^r y_{2n+1})}$$

Since $\{A^r x_{2n}\} = \{B^r x_{2n}\} \to z$, $\{S^r x_{2n+1}\} = \{T^r x_{2n+1}\} \to z$, $B^r B^r x_{2n} \to B^r z$, and $T^r B^r x_{2n} \to B^r z$. If we take limits in the above inequality we deduce the following

$$d(B^r z, z) \leq Z\beta^r \frac{d(B^r z, z)^2}{d(B^r z, B^r z) + d(B^r z, z)}$$
$$\leq Z\beta^r \frac{d(B^r z, z)^2}{3d(B^r z, z)}$$
$$= Z\beta^r \frac{d(B^r z, z)}{3}$$

Since $1 - \frac{Z\beta^r}{3} \neq 0$, the above inequality implies $d(B^r z, z) = 0$, that is, $B^r z = z$. Now assume S is r-continuous. Since the pair (S, A) is r-compatible of type (A), then by Proposition 9.2.9 we have, $S^r S^r x_{2n} \to S^r z$ and $A^r S^r x_{2n} \to S^r z$. If we consider $x = x_{2n+1}$ and $y = S^r x_{2n}$ in Definition 9.3.4, and proceed as in the above cases, then we obtain $S^r z = z$ for any $r \in \mathbb{N}$. Similarly, we can show that if the pair (S, A) is r-compatible of type (A) and A is r-continuous, then $A^r z = z$ for any $r \in \mathbb{N}$. It now follows that $A^r z = B^r z = S^r z = T^r z = z$. Thus, z is the common r-fixed point of the mappings A, B, S, T. Finally, we show uniqueness of the common r-fixed point. Suppose z and w are common r-fixed points of the mappings A, B, S, T, but, $z \neq w$. Since (T, S) is a higher-order $(\alpha, \beta, \gamma, \kappa, \delta, \mu)$- type contraction with respect to (A, B) we deduce the following

$$d(z, w) = d(T^r z, S^r w)$$
$$\leq Z\beta^r M(z, w)$$

where

$$M(z, w) = \frac{d(w, w) d(z, w)}{d(z, z) + d(w, z)} + 2 \frac{d(w, w) d(z, z)}{d(z, z) + d(w, z)}$$
$$+ 2 \frac{d(w, w) d(z, z)}{d(z, w) + d(w, z)} + \frac{d(z, w) d(z, w)}{d(z, z) + d(w, z)}$$

Proof of Theorem 9.3.5 Continued

Now observe that the upper bound on $M(z, w)$ is $\frac{23}{3} d(z, w)$. Thus, from the above inequality, we deduce that,

$$d(z, w) \leq \frac{23}{3} Z\beta^r d(z, w)$$

and since $1 - \frac{23}{3} Z\beta^r \neq 0$ we get that $d(z, w) = 0$, that is, $z = w$

Let $A = B$ in the previous Theorem, and let Z' be the modification on Z, where Z is given by Proposition 9.3.3, then we obtain the following

Corollary 9.3.6

Let (X, d) be a complete d-metric space, and A, S, T be three self maps of X. Suppose the following conditions hold

(a) For any $r \in \mathbb{N}$, $T^r(X) \subset A^r(X)$ and $S^r(X) \subset A^r(X)$

(b) The pairs (T, A) and (S, A) are r-compatible of type (A)

(c) (T, S) is a higher-order $(\alpha, \beta, \gamma, \kappa, \delta, \mu)$- type contraction with respect to A, that is,

$$d(T^r x, S^r y) \leq Z' \beta^r M'(x, y)$$

where

$$M'(x, y) = \frac{d(Ay, Sy)d(Ax, Ay)}{d(Ax, Tx) + d(Sy, Ax)} + \frac{d(Tx, Ax)d(Ty, Ay)}{d(Ax, Tx) + d(Sy, Ax)}$$
$$+ \frac{d(Ax, Sx)d(Sy, Ay)}{d(Ax, Tx) + d(Sy, Ax)} + \frac{d(Ax, Ay)d(Tx, Sy)}{d(Ax, Tx) + d(Ax, Sy)}$$
$$+ \frac{d(Ax, Tx)d(Ay, Sy)}{d(Ax, Ty) + d(Ax, Ty)} + \frac{d(Ax, Sx)d(Ay, Ty)}{d(Ax, Ay) + d(Ax, Ty)}$$

$Z' \geq 1$ is modified from Proposition 9.3.3, and $\beta \in [0, \frac{1}{6})$

If one of A, S, T is r-continuous, then A, S, T have a unique common r-fixed point in X

Let $A = B$ and $T = S$ in the previous Theorem, and let Z'' be the modification on Z, where Z is given by Proposition 9.3.3, then we obtain the following

Corollary 9.3.7

Let (X, d) be a complete d-metric space, and A, S be two self maps of X. Suppose the following conditions hold

(a) For any $r \in \mathbb{N}$, $S^r(X) \subset A^r(X)$

(b) The pair (S, A) is r-compatible of type (A)

(c) S is a higher-order $(\alpha, \beta, \gamma, \kappa, \delta, \mu)$- type contraction with respect to A, that is,

$$d(T^r x, S^r y) \leq Z'' \beta^r M''(x, y)$$

where

$$M''(x, y) = \frac{d(Ay, Sy)d(Ax, Ay)}{d(Ax, Sx) + d(Sy, Ax)} + \frac{d(Sx, Ax)d(Sy, Ay)}{d(Ax, Sx) + d(Sy, Ax)}$$
$$+ \frac{d(Ax, Sx)d(Sy, Ay)}{d(Ax, Sx) + d(Sy, Ax)} + \frac{d(Ax, Ay)d(Sx, Sy)}{d(Ax, Sx) + d(Ax, Sy)}$$
$$+ \frac{d(Ax, Sx)d(Ay, Sy)}{d(Ax, Sy) + d(Ax, Sy)} + \frac{d(Ax, Sx)d(Ay, Sy)}{d(Ax, Ay) + d(Ax, Sy)}$$

$Z'' \geq 1$ is modified from Proposition 9.3.3, and $\beta \in [0, \frac{1}{6})$

If one of A, S is r-continuous, then A, S have a unique common r-fixed point in X

Remark 9.3.8

Corollaries similar to Corollary 9.3.6 and Corollary 9.3.7 can be obtained in the following cases

(a) $A = B = I$ and $T = S$

(b) $A = B = I$

(c) $T = S$

9.4 Open Problem I

The open problem here concerns obtaining special cases of Theorem 9.3.5. In particular the problems are as follows

Corollary 9.4.1

State the Corollary arising from Theorem 9.3.5, when $A = B = I$ and $T = S$

Corollary 9.4.2

State the Corollary arising from Theorem 9.3.5, when $A = B = I$

Corollary 9.4.3

State the Corollary arising from Theorem 9.3.5, when $T = S$

9.5 Open Problem II

Definition 9.5.1

Let (X, d) be a d-metric space, and $A, B, S, T : X \mapsto X$ be four self mappings. We say (T, S) is a $(\alpha, \beta, \gamma, \kappa, \delta, \mu)$- type contraction with respect to (A, B) for some positive integers p, q, f, g if the following holds for all $x, y \in X$ and $k \in [0, \frac{1}{6})$

$$d(T^g x, S^f y) \leq k M(x, y)$$

where

$$M(x,y) = \frac{d(A^p y, S^f y) d(B^q x, A^p y)}{d(A^p x, T^g x) + d(S^f y, A^p x)} + \frac{d(T^g x, A^p x) d(T^g y, B^q y)}{d(A^p x, T^g x) + d(S^f y, A^p x)}$$
$$+ \frac{d(A^p x, S^f x) d(S^f y, A^p y)}{d(A^p x, T^g x) + d(S^f y, A^p x)} + \frac{d(B^q x, A^p y) d(T^g x, S^f y)}{d(B^q x, T^g x) + d(B^q x, S^f y)}$$
$$+ \frac{d(B^q x, T^g x) d(A^p y, S^f y)}{d(A^p x, T^g y) + d(B^q x, T^g y)} + \frac{d(A^p x, S^f x) d(B^q y, T^g y)}{d(B^q x, A^p y) + d(A^p x, T^g y)}$$

Now the higher-order version of the above reads as follows

Definition 9.5.2

Let (X, d) be a d-metric space, and $A, B, S, T : X \mapsto X$ be four self mappings. We say (T, S) is a higher-order $(\alpha, \beta, \gamma, \kappa, \delta, \mu)$- type contraction with respect to (A, B) for some positive integers p, q, f, g if the following holds for all $x, y \in X$, $0 \leq c_h < \frac{1}{6}$, $0 \leq h \leq r - 1$, and $r \in \mathbb{N}$

$$d(T^{rg}x, S^{rf}y) \leq \sum_{h=0}^{r-1} c_h T(x, y)$$

where

$$T(x,y) = \frac{d(A^{p(h+1)}y, S^{f(h+1)}y)d(B^{q(h+1)}x, A^{p(h+1)}y)}{d(A^{p(h+1)}x, T^{g(h+1)}x) + d(S^{f(h+1)}y, A^{p(h+1)}x)}$$
$$+ \frac{d(T^{g(h+1)}x, A^{p(h+1)}x)d(T^{g(h+1)}y, B^{q(h+1)}y)}{d(A^{p(h+1)}x, T^{g(h+1)}x) + d(S^{f(h+1)}y, A^{p(h+1)}x)}$$
$$+ \frac{d(A^{p(h+1)}x, S^{f(h+1)}x)d(S^{f(h+1)}y, A^{p(h+1)}y)}{d(A^{p(h+1)}x, T^{g(h+1)}x) + d(S^{f(h+1)}y, A^{p(h+1)}x)}$$
$$+ \frac{d(B^{q(h+1)}x, A^{p(h+1)}y)d(T^{g(h+1)}x, S^{f(h+1)}y)}{d(B^{q(h+1)}x, T^{g(h+1)}x) + d(B^{q(h+1)}x, S^{f(h+1)}y)}$$
$$+ \frac{d(B^{q(h+1)}x, T^{g(h+1)}x)d(A^{p(h+1)}y, S^{f(h+1)}y)}{d(A^{p(h+1)}x, T^{g(h+1)}y) + d(B^{q(h+1)}x, T^{g(h+1)}y)}$$
$$+ \frac{d(A^{p(h+1)}x, S^{f(h+1)}x)d(B^{q(h+1)}y, T^{g(h+1)}y)}{d(B^{q(h+1)}x, A^{p(h+1)}y) + d(A^{p(h+1)}x, T^{g(h+1)}y)}$$

Proposition 9.5.3

Let A, B, S, T be four self maps of a d-metric space (X, d), where (T, S) is a higher-order $(\alpha, \beta, \gamma, \kappa, \delta, \mu)$- type contraction with respect to (A, B) for some positive integers p, q, f, g. Put

$$M(x,y) = \frac{d(A^p y, S^f y)d(B^q x, A^p y)}{d(A^p x, T^g x) + d(S^f y, A^p x)} + \frac{d(T^g x, A^p x)d(T^g y, B^q y)}{d(A^p x, T^g x) + d(S^f y, A^p x)}$$
$$+ \frac{d(A^p x, S^f x)d(S^f y, A^p y)}{d(A^p x, T^g x) + d(S^f y, A^p x)} + \frac{d(B^q x, A^p y)d(T^g x, S^f y)}{d(B^q x, T^g x) + d(B^q x, S^f y)}$$
$$+ \frac{d(B^q x, T^g x)d(A^p y, S^f y)}{d(A^p x, T^g y) + d(B^q x, T^g y)} + \frac{d(A^p x, S^f x)d(B^q y, T^g y)}{d(B^q x, A^p y) + d(A^p x, T^g y)}$$

Now for every pair $x \neq y$, define

$$Z := Z(x, y) = \max_{0 \leq v \leq r-1} \beta^{-v} \frac{d(T^{vg}x, S^{vf}y)}{M(x, y)}$$

then

$$Z = \max_{n \in \mathbb{N} \cup \{0\}} \beta^{-n} \frac{d(T^{ng}x, S^{nf}y)}{M(x, y)}$$

where $\beta \in [0, \frac{1}{6})$

Now we have the following alternate characterization of the higher-order $(\alpha, \beta, \gamma, \kappa, \delta, \mu)$- type contraction for some positive integers p, q, f, g

CHAPTER 9. COMMON R-FIXED POINT THEOREMS FOR MAPPINGS IN DISLOCATED METRIC SPACE UNDER R-COMPATIBILITY OF TYPE (A)

> **Definition 9.5.4**
>
> Let A, B, S, T be four self maps of a d-metric space (X, d), we say (T, S) is a higher-order $(\alpha, \beta, \gamma, \kappa, \delta, \mu)$- type contraction with respect to (A, B) for some positive integers p, q, f, g if the following holds for all $x, y \in X$ and $r \in \mathbb{N}$
>
> $$d(T^{rg}x, S^{rf}y) \leq Z\beta^r M(x,y)$$
>
> where
>
> $$M(x,y) = \frac{d(A^p y, S^f y) d(B^q x, A^p y)}{d(A^p x, T^g x) + d(S^f y, A^p x)} + \frac{d(T^g x, A^p x) d(T^g y, B^q y)}{d(A^p x, T^g x) + d(S^f y, A^p x)}$$
> $$+ \frac{d(A^p x, S^f x) d(S^f y, A^p y)}{d(A^p x, T^g x) + d(S^f y, A^p x)} + \frac{d(B^q x, A^p y) d(T^g x, S^f y)}{d(B^q x, T^g x) + d(B^q x, S^f y)}$$
> $$+ \frac{d(B^q x, T^g x) d(A^p y, S^f y)}{d(A^p x, T^g y) + d(B^q x, T^g y)} + \frac{d(A^p x, S^f x) d(B^q y, T^g y)}{d(B^q x, A^p y) + d(A^p x, T^g y)}$$
>
> $Z \geq 1$ is given by the previous Proposition, and $\beta \in [0, \frac{1}{6})$

Now the open problem is to prove the following

> **Theorem 9.5.5**
>
> Let (X, d) be a complete d-metric space, and A, B, S, T be four self maps of X. Suppose the following conditions hold
>
> (a) For any $r \in \mathbb{N}$, $T^{rg}(X) \subset A^{rp}(X)$ and $S^{rf}(X) \subset B^{rq}(X)$ for some positive integers p, q, f, g
>
> (b) The pairs (T^g, B^q) and (S^f, A^p) are r-compatible of type (A) for some positive integers p, q, f, g
>
> (c) (T, S) is a higher-order $(\alpha, \beta, \gamma, \kappa, \delta, \mu)$- type contraction with respect to (A, B) for some positive integers p, q, f, g
>
> If one of A^p, B^q, S^f, T^g is r-continuous for some positive integers p, q, f, g, then A^p, B^q, S^f, T^g have a unique common r-fixed point in X

Bibliography

[1] Clement Ampadu, Fixed Point Theory for Higher-Order Mappings. ISBN: 5800118959925, lulu.com, 2016

[2] Fei He, Common fixed point of four self maps on dislocated metric spaces, J. Nonlinear Sci. Appl. 8 (2015), 301-308

[3] P. Hitzler, A. K. Seda, Dislocated Topologies, J . Electr. Engin., 51 (2000), 3-7

[4] I. Altun, G. Durmaz, Weak partial metric spaces and some fixed point results, Appl. Gen. Topol., 13 (2012), 179-191

[5] G. Durmaz, O. Acar, I. Altun, Some fixed point results on weak partial metric spaces, Filomat, 27 (2013), 317-326

[6] Dinesh Panthi, Kumar Subedi, Some Common Fixed Point Theorems for Four Mappings in Dislocated Metric Space, Advances in Pure Mathematics, 2016, 6, 695-712

[7] Jungck, G. and Rhoades, B.E. (1998) Fixed Points for Set Valued Functions without Continuity. Indian Journal of Pure and Applied Mathematics , 29, 227-238

[8] Amri, M. and El Moutawakil, D. (2002) Some New Common Fixed Point Theorems under Strict Contractive Conditions. Journal of Mathematical Analysis and Applications , 270, 181-188

[9] Sintunavarat, W. and Kumam, P. (2011) Common Fixed Points for a Pair of Weakly Compatible Maps in Fuzzy Metric Spaces. Journal of Applied Mathematics , 2011, 1-14

[10] K. Wadhwa, H. Dubey and R. Jain, Impact of E. A. Like property on common fixed point theorems in fuzzy metric spaces. J. Adv. Stud. Topology 3 (1) (2012), 52-59

[11] Kastriot Zoto, Ilir Vardhami, Jani Dine and Arben Isufati, COMMON FIXED POINTS IN b-DISLOCATED METRIC SPACES USING (E.A) PROPERTY, BULLETIN MATHÉMATIQUE, Vol.40(LXVI) No. 1 2014 (15-27)

[12] R. Shrivastava, Z. K. Ansari and M. Sharma, Some results on Fixed Points in Dislocated and Dislocated Quasi-Metric Spaces, Journal of Advanced Studies in Topology, Vol. 3, No.1, (2012)

[13] M. A. Kutbi, M. Arshad, J. Ahmad, A. Azam, Generalized common fixed point results with applications, Abstract and Applied Analysis, volume 2014, article ID 363925, 7 pages

[14] N. Hussain, J. R. Roshan, V. Parvaneh and M. Abbas, Common fixed point results for weak contractive mappings in ordered b-dislocated metric spaces with applications, Journal of inequalities and Applications, 1/486, (2013)

[15] I. Beg and M. Abbas, Coincidence and common fixed points of noncompatible maps. J. Appl. Math. Inform. 29(3-4), 743-752 (2011)

[16] G. Jungck, Compatible mappings and common fixed points, Int. J. Math. Math. Sci., (1986), 771-779

[17] Dinesh Panthi and P. Sumati Kumari, Common Fixed Point Theorems for Mappings of Compatible Type(A) in Dislocated Metric Space, Nepal Journal of Science and Technology Vol. 16, No 1 (2015) 79-86

www.ingramcontent.com/pod-product-compliance
Lightning Source LLC
Chambersburg PA
CBHW051054180526
45172CB00002B/636